The Compleat Cockroach

THE
COMPLEAT
COCKROACH

A Comprehensive Guide to the Most Despised (and Least Understood) Creature on Earth

DAVID GEORGE GORDON

TEN SPEED PRESS
Berkeley, California

1🔟

Ten Speed Press
P.O. Box 7123
Berkeley, California 94707

Distributed in Australia by E. J. Dwyer Pty. Ltd., in Canada by Publishers Group West, in New Zealand by Tandem Press, in South Africa by Real Books, in Singapore and Malaysia by Berkeley Books, and in the United Kingdom and Europe by Airlift Books.

Cover design by Big Fish Books, San Francisco
Interior design by Catherine Jacobes
Illustrations by Jim Hays

Image on title page courtesy of Dennis Kunkel, University of Hawaii.; David Letteroach on page 50 created by Michael Bohdan of The Pest Shop, Inc. in Plano, Texas; Photographs on pages 54, 95 by Scott Stenjem of Stenjem Photography; Hand-puppet on page 59 courtesy of Folkmanis, Inc., Emeryville, California; Roach crossing sign on page 84 from Atlas Screenprinting of Gainesville, Florida; Roachill Downs photos on pages 85 and 86 courtesy of Arwin Provonsha, Department of Entomology, Purdue University; Robo-roach photo on page 87 courtesy of Fred Delcomyn, Department of Entomology, University of Illinois at Urbana-Champaign; Image on page 118 is by Maria Sibylla Merian (German, 1647–1717). *Dissertations in Insect Generations and Metamorphosis in Surinam*, 1719. Hand-colored engraving from volume of seventy-two engravings, second edition. The National Museum of Women in the Arts. Gift of Wallace and Wilhelmina Holladay. A detail from this work also appears on the cover; "Twelfth Station," by Manuel Ocampo, on page 119, courtesy of Anna Nosei Gallery, New York; Roachart on page 120 created by Dick Webb and photographed by Bob Kelly; Still from *Twilight of the Cockroaches* © 1987 TYP Productions, Inc., Kitty Films, Inc. Courtesy of Streamline Pictures, Santa Monica, California; All excerpts from *archy & mehitabel* courtesy of Amereon, Ltd.; Excerpt from "The Old Gumbie Cat" in OLD POSSUM'S BOOK OF PRACTICAL CATS, copyright 1939 by T. S. Eliot and renewed 1967 by Esme Valerie Eliot, reprinted by permission of Harcourt Brace & Company.

Library of Congress Cataloging-in-Publication Data
Gordon, David G. (David George), 1950–
    The compleat cockroach: a comprehensive guide to the most
despised (and least understood) creature on earth / [David George Gordon].
      p. cm.
    Includes bibliographical references (p.   ) and index.
    ISBN 0-89815-853-2
    1. Cockroaches.   I. Tiltle.
QL505.5.G67   1996   96–2736
                                                       CIP

First printing, 1996
Printed in Canada

2 3 4 5 6 7 8 9 10 — 00 99 98 97

For Rosalie, who first encouraged me to write,
and who taught me to seek out
the good in everyone.

Acknowledgments

THIS BOOK CONTAINS THE COLLECTED wisdom of several hundred individuals—entomologists, pest control specialists, psychologists, filmmakers, novelists, historians, fine and folk artists, and a few of my close friends.

I am especially grateful to Dr. Ivan Huber of Farleigh Dickinson University, who counseled me on matters entomological, and to Dr. Louis Roth of Harvard University's Museum of Comparative Zoology, who patiently answered my many mundane questions. Other experts generously contributed their expertise: William Bell of the University of Kansas, May Berenbaum of the University of Illinois at Champaign-Urbana, Donald Cochran and Mary Ross of Virginia Polytechnic Institute, Betty Faber with the Liberty Science Center in Jersey City, Dr. Philippe Grandcolas at the University de Rennes, Phil Koehler at the University of Florida, Alan Moore at the University of Kentucky, Michael Rust at the University of California at Riverside, Coby Schal and Christine Nalepa at North Carolina State University, Barbara Stay at the University of Iowa, Dan Suiter at Purdue University, and Susan Silanger with the U.S. Fish and Wildlife Service's Caribbean field office. I thank them for their help.

All the contributors of cultural information and care-and-feeding tips are, alas, too numerous to acknowledge on this page. Among my many helpers were Kraig Anderson, Jerry Beck, Judith Bows, Sharon Collman, Stuart Fullerton, Stuart Gordon, Mark Kausler, Steve Kutcher, Art Evans, Richard Leskosky, Gail Manning, Stefan LeTirant, Mary McCoy, Randy Morgan, Dave Nickel, Steve

Prchal, Stephen Puchalski, Gordon Ramel, Shannon Williams, John Waters, and Elizabeth Watson. I am deeply indebted to everyone who chipped in.

I am thankful for Mariah Bear, a superb editor with unlimited love for this book's six-legged subjects, and Jim Hays, friend and illustrator, who can even make the leg hairs of a cockroach look good. *The Compleat Cockroach* would not exist without the vision of Kirsty Melville, the design wizardry of Catherine Jacobes, the lovely cover from John Miller of Big Fish, the tenacity of my agent, Anne Depue, and the unflagging faith of my wife, Mari Mullen.

Finally, I wish to thank all cockroaches—for reminding us that we are not superior, nor are we ever really alone.

Contents

INTRODUCTION / xi

COCKROACH BASICS

What Is a Cockroach? / 3

Where Are (or Aren't) They Found? / 19

A 340-Million-Year History / 33

How They Affect Our Lives / 45

SEX, FOOD, AND DEATH

This Is How It All Begins... / 65

Cockroach Transportation / 83

Gastronomy / 97

Wasps, Cats, and Other Perils / 105

WHEN HUMANS AND COCKROACHES MEET

Cockroaches in Human Culture / 117

Cockroaches as Pets and Prizes / 135

Can We Control Cockroaches? / 149

RESOURCES / 171

INDEX / 175

a good many

failures are happy

because they don't

realize it many a

cockroach believes

himself as beautiful

as a butterfly

have a heart o have

a heart and

let them dream on

—"ARCHYGRAMS" from
archy's life of mehitabel

Introduction

IN THE SUMMER OF 1995, I received a package from Federal Express. Inside it was a round plastic container, like the ones in which cottage cheese is sold. Within this container, comfortably padded with vermiculite, were five leathery, brown insects, the largest of which was a good four inches long.

These were not your ordinary cockroaches. They were Madagascan hissing cockroaches, adults in the prime of their lives with shiny black legs and wobbly antennae. I was enthralled.

Along with this carton of hissing cockroaches was a note from my friend Kraig Anderson, founder of Spineless Wonders, a traveling insect zoo in Minneapolis. "Live with them and know their ways," the note read.

Since that eventful day, I *have* lived with cockroaches—not just with the five hissers (which I now keep in a fifteen-gallon terrarium in my office), but with cockroach kitchen magnets, T-shirts, and videotapes that I've also acquired. Not to mention the trade publications about pest control, the rubber roaches in four different sizes, the hand puppet, the enameled "Roach Crossing" signs. Forget about the many versions of *La Cucaracha* that I heard, or the many awful horror movies about bugs that I screened in my home.

I spent countless hours in the University of Washington's Natural Sciences Library searching for articles in the various entomological annals, proceedings, journals, and reports, some dating back to the 1820s. Photocopying each informative scrap of paper, stuffing the copies into file folders for later use— at times I felt more like a clerk than a nature writer.

At the end of this phase in my project, I weighed the volume of paper I'd assembled and found that it was a full forty pounds. That's not counting books—some treasures like the *Biotic Associations of Cockroaches* by Roth and Willis, or P. B. Cornwell's *The Cockroach: A Laboratory Insect and Industrial Pest.* I filled two shelves with titles like these.

I spoke with cockroach scholars around the world, interviewing field researchers, pest control operators, professors, just about anyone with an intimate knowledge of my six-legged subjects. I made a nuisance of myself with several Internet news groups, yet managed to make contact with additional experts in Australia, South Africa, and France. I bought the soundtrack to *Hairspray* and a few other CDs with cockroach-related songs. I started wearing a small, brass cockroach charm from Mexico.

By now, my Madagascan hissers—Estelle, Richard, Bonnie, Sally, and Louis—had molted their skins several times. I'd gotten used to their hissing, and my daughter Julia and I had become rather adept at handling them. Having compiled all the raw data I could absorb, I was now ready to write *The Compleat Cockroach: A Comprehensive Guide to the Most Despised (and Least Understood) Creature on Earth.*

The title makes three rather extraordinary claims. The first is that the work is complete—a statement that no one should seriously make during this, the early years of the Information Age. Of course this book isn't complete. It *is*, however, compleat—as thorough and well ordered as any first edition of an all-purpose reference book will ever get. If you know something about cockroaches that isn't already in this volume, by all means write me care of the publisher, and I'll make an effort to include it in a subsequent edition.

Is the cockroach the most despised animal on our planet? Certainly. P. B. Cornwell said so in the first sentence of his book: "The cockroach is probably the most obnoxious insect known to man."

Cornwell's view was apparently shared by Glenn W. Herrick, professor of entomology at Cornell University for twenty-six years. "Cockroaches are exceedingly annoying from the mere fact of their presence and their disgusting proneness to get into things," wrote the distinguished author of *Insects Injurious to the Household and Annoying to Man*, which was published in 1914.

If you're still unconvinced, seek out the March 1982 issue of *Pest Control*, in which the following paragraph appeared:

> In 1981, when the U.S. Fish and Wildlife Service polled 3,107 adults, they learned that cockroaches were America's least favorite animals, followed by mosquitoes, rats, wasps, rattlesnakes, and bats. The study also revealed that Americans knew little about wildlife and related issues: a majority of those surveyed thought spiders had ten legs and that an iguana is an insect.

Or find the article in the July 1994 *Scientific American*:

> "U.S. consumers bought $240 million worth of anti-cockroach toxins last year."

Ask yourself how *you* feel about cockroaches. Do you shiver at the mere mention of their name? Could you kill one with your bare hand? Would you rather not be talking about this?

Now what about this business of being the least-understood creature on earth? I told you about all of those journal articles, and indeed, there are literally thousands of them. There are also more than a dozen thick textbooks on cockroach anatomy, physiology, and behavior. But the sad truth is that nearly all of those works are centered around a handful of species that are considered pests. Of these all-too-familiar critters—the American, German, Oriental, brownbanded, and smokeybrown cockroaches—we know plenty. We know considerably less about another twenty species that have also been labeled domestic pests. And we know nearly nothing about the vast majority of nonpest cockroaches, some 3,500 species by most recent count.

Until recently, these wild cockroaches have been of little importance to most of us. They, in turn, have been perfectly content to pursue meaningful lives in the forests of the tropics. They've done fine without any contact whatsoever with humankind, and they'd probably like to keep it that way. But after centuries of neglect, a few of these cockroaches are being examined in depth by people striving to understand the world's forests, and the secrets of these essential ecosystems.

In the process, we are beginning to regard the cockroach in a new light. Instead of an accursed nuisance, we are seeing a wizened old soul—one whose ancestors were around when the continents formed, who witnessed the emergence and disappearance of the dinosaurs, and who watched an agile, chimpanzee-like primate become *Homo sapiens*, alternately a domestic cockroach's worst enemy and best friend.

They're intelligent, hardworking, and fastidious groomers—who wouldn't want a few cockroaches around? Of course, even one so enamored as I with these animals can see that cockroaches, in particular the domestic pest species, can be a real pain. Years ago, when I lived in Chicago, cockroach infestations were practically written into every apartment rental agreement I signed. Several million Americans suffer from cockroach-related allergies, and, while it's never been proven, cockroaches probably play a part in the transmission of many diseases.

To appease those seeking relief from pest populations, I've incorporated the latest information on cockroach control into this book. This material is presented in an escalating order of severity, with the least-toxic alternatives presented first. To ensure the continued survival of the many thousands of wild cockroach species, as well as that of our own species, readers are encouraged to select the most environmentally sound pest-control alternatives—those with the least potential for harming the health of humans and other animals—and to launch hardcore chemical weapons only as a last resort.

By reading my words and those of the many scientists and observers upon which I've drawn, you will better understand and appreciate my friend *La Cucaracha*, the most despised creature on our planet. For only then will you be in a position to forge a meaningful and lasting peace with this animal—one of the oldest and most successful beings on earth.

Cockroach Basics

What Is a Cockroach?

COCKROACHES HAVE FLATTENED, oval-shaped bodies, down-turned heads, long antennae, and, in most species, two pairs of wings. They are exothermic (cold-blooded) invertebrates, without backbones or other internal supports made of cartilage or bone. They belong to one of the oldest, largest, and most successful groups of animals on Earth—the insects.

Insects are the dominant form of life on our planet, with a strong presence in the fossil record that dates back some 400 million years. They occupy an impressive array of niches, from scorching deserts and boiling hot springs to snowcapped mountains and frigid fjords. As plant pollinators, prey items, and nutrient recyclers, they are essential components of coexistence. If all insects were to suddenly disappear, hundreds of thousands of species of vertebrates and plants would soon perish.

Insects are also the only animals in direct competition with humankind for the world's food resources. As agricultural pests, urban nuisances, and carriers of malaria and other life-threatening diseases, they make formidable opponents.

As of this writing, entomologists (scientists specializing in the study of insects) have identified over one million species of insects. It's thought that this figure represents a small fraction of the real number of living insect species, the vast majority of which have yet to be captured, catalogued, and named.

Key characteristics

Cockroaches and all other insects wear their skeletons on the outside. This skeleton is made of chitin (pronounced *kye-tin*), a durable, polysaccharide "shell" no more than a hundred microns thick—about the width of a human hair—in any spot. It covers the entire body, even the eyes, and is jointed like a medieval knight's armor to facilitate movement of the limbs.

Unlike a tin suit, a cockroach's chitinous covering is remarkably lightweight. With more muscles than humans (792 distinct muscles in humans; 900 in grasshoppers), all insects benefit from the superior leverage of their internally attached muscles to pull, on average, more than twenty times their own weight.

The bodies of cockroaches and other insects are composed of three segments: head, thorax, and abdomen. In nearly all species, the head is wholly concealed by the shieldlike pronotum, a convex plate of chitin that is often distinctively patterned or colored. The posterior end of this plate also covers the points of attachment for two pairs of wings.

Not every kind of cockroach is winged. Some species are entirely wingless, while others are graced with small, vestigial nibs. In several species, only the members of one sex possess these specialized appendages for flight. A few more have wings but seldom use them, spreading these solely when breaking a fall.

Directly below the pronotum are the points of attachment for three pairs of hairy and heavily spined legs. At the end of each leg is a foot bearing two large claws. These can be used like grappling hooks for hauling their bodies up vertical surfaces. To facilitate fast movement over slick surfaces, cockroach feet and forelegs also have special adhesive pads.

A cockroach's abdomen typically comprises more than half of its bulk. The softest, most flexible part of the body, it is protected by two sets of armor plates—upper ones called *tergites*, and lower ones called *sternites*. These plates can be used to sex specimens, as females have only seven visible sternites when viewed from below, while males have nine.

Arranged in rows on either side of the abdomen are small holes (called spiracles), each attached to an air tube, through which these animals breathe. Cockroaches and their insect kin are the only animals on earth that rely on this kind of breathing apparatus.

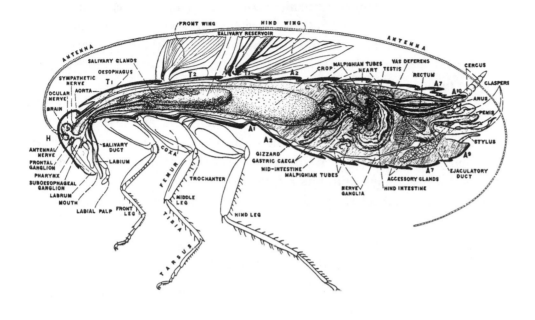

Cockroaches classified

On the basis of certain shared traits, entomologists have divided all living insects into twenty-six categories, known as orders. Within each of these orders are several smaller subsets—suborders, complexes, families, genera, species, and subspecies. These subsets have been created to assist taxonomists (people who specialize in classifying plants and other creatures) in establishing relationships among the various members of the animal kingdom. For most students of cockroach biology, terms such as *family* or *complex* are less meaningful than *genus* and *species*—the scientific "surname" and "first name" of an insect.

All cockroaches have been assigned to the order Blattaria, named for the *blattae*, the domestic pest insects of the ancient Greeks. Members of this order (the *blattarians*, as they are commonly called) share certain key characteristics. All have thick, leathery forewings, grasshopper-like mouth parts designed for chewing, and simplified life cycles that lack larval (caterpillar) and pupal (cocoon) stages of growth.

Another more common trait among all cockroaches is the egg capsule, or ootheca (pronounced *oh-a-theek-a*)—a small, hard-shelled "purse" in which females deposit their eggs. Termites (of the order Isoptera) and mantids (order Mantoidea) are the only other insects with carrying cases for their eggs. Because of this and a few other similarities, entomologists once put all cockroaches, termites, and mantids in one order—the Dictyoptera or "net-winged" insects. Some taxonomists also included grasshoppers and crickets in this melange, cramming all of these animals into the order Orthoptera. Neither Dictyoptera nor Orthoptera are considered accurate taxonomic units today. However, it is not unusual to find references to the orthoptera—the grasshoppers, crickets, cockroaches, and their allies—in both scientific and popularized texts on insects.

Family affairs

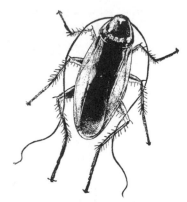

Five cockroach families have been identified within the order Blattaria: Blattidae, Blaberidae, Blatellidae, Polyphagidae, and Cryptocercidae. Within these families is considerable diversity—in the ranges of shapes, sizes, and colors. While introducing each member of the five families is clearly beyond the scope of this book, a few representatives of the various types are provided here.

BLATTIDAE

The American cockroach (*Periplaneta americana*), or palmetto bug, as it is euphemistically called, is the most familiar species of the Blattidae. This medium-sized insect and its close cousins, the Australian (*P. australasiae*) and brown (*P. brunnea*), are now fairly evenly distributed throughout the warmer parts of the world. *P. americana* is frequently used as a laboratory experimental animal because of its abundance, size, and the ease with which it is bred in laboratory cultures.

BLABERIDAE

"Cockroach of the divine face" is what natives of Costa Rica call the giant cockroach whose scientific name is *Blaberus giganteus*, one of several large members of the Blaberidae. Common in caves, rock crevices, and hollow trees, this species feeds on rotting wood, bat guano, decomposing plants, dead insects, and just about anything else that drops into these dank recesses. Captive-bred giant cockroaches are popular attractions at insect zoos. They also make nice pets.

BLATTELLIDAE

A member of the Blattellidae, the field cockroach (scientfic name *Blattella vaga*) ordinarily inhabits irrigated fields and heavily mulched landscapes. However, this small moisture-loving resident of the American Southwest has been frequently observed entering houses in large numbers during dry seasons. Here, it shows little of the cockroach's characteristic aversion to bright light, wandering freely and openly in the glare of incandescent and fluorescent light fixtures.

POLYPHAGIDAE

Desert cockroaches (*Arenivaga sp.*) of the family Polyphagidae are well adapted for life in inhospitable surroundings. During the heat of the day, they burrow in the top few inches of sand, emerging to feed and seek mates only after the sun goes down. By sticking out special "tongues" (more accurately, their *hypopharyngeal bladders*), *Arenivaga* and other Polyphagidac can absorb water vapor directly from the air.

CRYPTOCERCIDAE

One of four surviving species of Cryptocercidae, the brown-hooded wood cockroach (*Cryptocercus punctulatus*) lives in decaying logs of remnant old-growth forests that once hugged the Atlantic coast from New York to Georgia. The dwindling habitat of this species and its shortage of relatives—C. *primarius* in China, *C. relictus* in Russia and eastern Manchuria, and a recently described third species, *C. clevelandi* from the western U.S.—may be signals that after millions of years, the Cryptocercidae family is approaching an evolutionary dead end.

As you read on, deeper into the world of blattarians, you will find more thorough descriptions of these and other cockroach species can be found throughout this book.

What's in a name?

Scientific names often contain clues about cockroach identity. Many of these monikers identify their bearer's presumed point of origin—for instance, *aegyptiana* (from Egypt) or *lapponica* (from Lapland). Several of these place names (including *germanica* and *americana*, both assigned in the mid-eighteenth century by the Swedish naturalist Carolus Linnaeus) have been erroneously applied to specimens taken far from their native shores.

Scientific names are also selected to honor scientists who have contributed significantly to the study of the Blattaria. A fine example is the genus *Miriamrothschildia*, dedicated to the Honorable Dr. Miriam Rothschild, biologist, conservationist, and Commander of the Order of the British Empire. Other names, such as *gigantea* (giant), *lutea* (yellowish), or *emarginata* (notched), are chosen to call attention to a cockroach's distinguishing physical traits.

Roach Perot, one of the "Best-Dressed Cockroaches" from Plano, Texas (see p. 26)

SCIENTIFIC NAME	MEANING
Diploptera punctata	Double wings covered with small holes (a unique characteristic of this genus)
Nauphoeta cinerea	Ash-colored seafarer (a familiar figure on sailing ships in tropical seas)
Periplaneta brunnea	Brown wandering star (a testimony to its near-universal distribution)
Blaberus craniifer	Noxious skull-bearer (because of the skull-like splotch on its back)
Gromphadorhina portentosa	Portentous sow's snout (why? I don't know)

While common names of insects change from place to place, scientific names change from time to time. After being assigned to a new branch on the blattarian family tree, a cockroach may be given a new generic name, species name, or both. An extreme example of such name changing is the German cockroach, which has been rechristened six times since 1767:

NAME	IN USE
Blatta germanica	1767-1792
Blatta obliquata	1793-1834
Ectobius germanicus	1835-1863
Phyllodromia bivittata	1864
Phyllodromia germanica	1865-1875
Ischnoptera bivittata	1876-1902
Blattella germanica	1903-present

Cucaracha, cockarooch, cockroach

You won't find "cockroach" in the Bible or in any of Shakespeare's works, for the word is much too young. Ancient Romans called the heat- and moisture-loving bugs in their public baths the *lucifugia*, a blanket term for cockroaches, rats, and other "light-fleeing" nocturnal pests. They continued to be known by this name or by that of the Greeks—the *blattae*—for several centuries.

"The German Cockroach, *Blatella germanica* ('small'+*Blatta*) is also known by a variety of names: the most common is 'steam-fly' because of its association with that type of environment; 'steam-bug' was apparently at one time a local name in Lancashire and 'shiner' around Aldershot. 'Water bug' is a common name used for cockroaches throughout the United States, 'Yankee settler' in Nova Scotia, and 'Croton bug' in the Eastern United States..."

—P.B. Cornwell, *The Cockroach* (1968)

The word cockroach became part of the English language in 1624—eight years after Shakespeare's death. It was in this year that Captain John Smith of the Virginia Colony wrote about "a certaine India Bug, called by the Spaniards a Cacarootch, the which creeping into Chests they eat and defile with their ill-scented dung." Smith no doubt misheard his Spanish contemporaries, who called these insects *cucarachas*—a name that combined the diminutive form of *cuco* or *cuca* (caterpillar) with *acha* (mean or contemptible).

By the mid-eighteenth century, Smith's word had become "cock-roche." It was later transformed into cock-roach, the spelling used by Charles Darwin in his 1859 edition of *The Origin of Species*. Around 1900, Americans shortened this already truncated word into "roach"—an etymological twist that irked at least one scholar of the Blattaria, Robert W. C. Shelford of Cambridge, England. In his 1917 account of *A Naturalist in Borneo*, Shelford wrote that the word roach "is good Anglo-Saxon for a species of fish"—*Rutilus rutilus*, a silvery European relative of the carp—so "the use of this word for an insect is objectionable."

Who knows what Shelford would've thought of the word's more recent association with be-bopping jazz drummer Max Roach? Or with the word's slang meaning—the butt of a marijuana cigarette? This second term, from the jargon of the Harlem hipster, was introduced to the public at large by Meyer Berger, a contributor to the March 12, 1938 issue of *The New Yorker*. In his piece *Tea for a Viper*, Berger shared his firsthand impressions of a pot party in progress. He told how "sticks," "reefers," "tea," "gyves," and "gauge or goofy butts" were all slang for the same thing—marijuana cigarettes.

"A pinched-off smoke, or stub, is a roach," Berger reported. Undoubtedly this fitting word for the brownish, bug-sized remnant was in use for many decades before Berger's trip to Harlem. It remains in use today, thinly disguised in many popular song lyrics and titles.

Incidentally, those who frequent the racetrack have been known to call a tired old nag a roach as well although, as we shall see, roaches can actually be quite speedy.

In sum, then, though I use the word roach, I have the greatest respect for Shelford, and I hope there are no readers who will misconstrue my meaning.

Basic blattarian

Cockroaches have changed little over the aeons. Like the designers of Volkswagen Beetles, they have stuck with what appears to be a winning formula, to which they've added countless refinements over time. The result of all this fine-tuning? An inconspicuous but well-equipped creature with alien features and abilities far beyond our own.

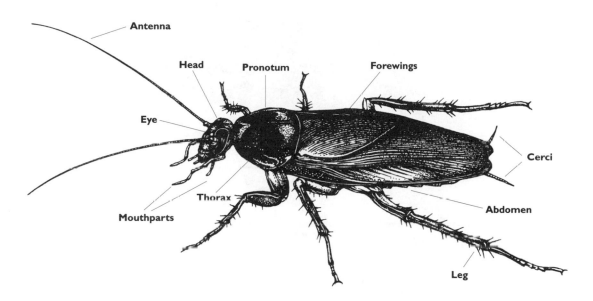

HEAD:

Antennae: long tubes consisting of as many as 130 segments, each outfitted with one or several receptors for temperature, motion, and scent.

Eyes: two types—compound eyes, each containing two thousand octagonal lenses, plus (in most species) simple eyespots without lenses, designed for detecting light and darkness.

Mouthparts: mandibles for chewing from side to side, armed with strong chitinous teeth; maxillae with tiny bristles for grooming antennae and legs. Finger-like maxillary palps and labial palps prod each bite of food, testing its edibility before actual ingestion.

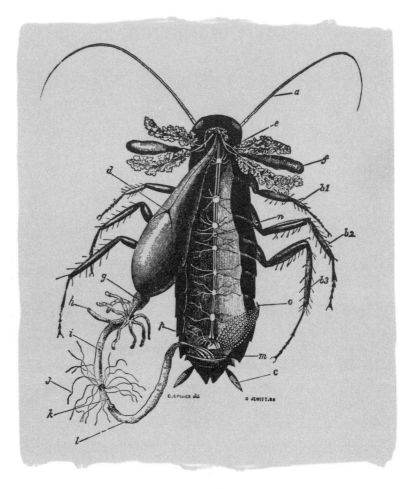

THORAX:

Legs: engineering marvels in miniatures, featuring three kinds of connections: a hinge where leg and body meet, a ball-and-socket joint attaching the tibia (shin) to the tarsus (foot), and knuckle joints at both "knees" (the junctures of the tibia and femur and the trochanter and coxa).

Ears: called *subgenual organs*, and located in each knee joint to detect the faintest of airborne sounds—even the footfalls of other roaches.

Wings: thick forewings concealing a more delicate, membranous pair of hindwings; a wingspread of three or four inches is not unusual.

ABDOMEN:

Spiracles: eight paired openings to the cockroach's tracheae; two additional pairs are situated on the thorax.

Cerci: jointed tail prongs with pressure-sensitive hairs that alert the roach of impending danger, bypassing the brain to flash warning messages directly to the legs.

Genitalia: (in males) three talonlike hooks, the longest of which is designed to clamp on to the female's abdominal tip—once this connection is made, several smaller hooks are attached, forming an unbreakable bond between the two animals; (in females) a slim orifice with internal hitching posts for the male's hooks to grab hold.

ENTIRE BODY:

Cuticle (or body wall): made of solid, seamless chitin, and protected by a waterproof veneer of oils and wax; keeps body moisture in and dust and disease-carrying organisms out. Cockroach cuticles come in an array of colors, from black or chocolate to yellow or bright green.

COCKROACH INNARDS:

Few animals are as well packaged as cockroaches. Neatly crammed into the unyielding exoskeleton is an array of internal systems. Highly miniaturized but no less efficient than our own, they provide the necessary respiratory, circulatory, digestive, excretory, reproductive, and sensory functions. Detailed descriptions of these systems can be found in *The American Cockroach* and *Cockroaches as Models for Neurobiology Applications in Biomedical Research*.

RESPIRATION AND CIRCULATION:

A cockroach's heart is contained in the abdominal segment. A thin tube pumps pigmentless blood through what is basically an open system, devoid of veins or arteries. This fluid courses through the body cavity, bathing internal tissues with nutrients from the digestive system and removing waste materials, which are conveyed to the excretory organs.

Metabolic oxygen is obtained from a convoluted system of tiny tubes (tracheoles), which are connected by slightly larger tubes (tracheae) attached to

Thanks to its two autonomous brains, a cockroach can remain active after its head is removed. Precautions must be taken to keep the decapitated subject from bleeding to death. Once this is done, however, the insect can live for several weeks before eventually dying of starvation.

the spiracles. Air enters the spiracles by diffusion, assisted by the contraction and relaxation of the roach's abdominal muscles. Thissounds complicated, and it is, but it's a lot simpler than our own breathing and circulation.

DIGESTION AND EXCRETION:

Once food has passed beyond a cockroach's mouthparts, it enters the cibarium for crushing. Then it travels backward into the salivarium (no, I'm not making this up), and mixes with spit. Thus readied for digestion, the pre-processed food slides into the oesophagus, then on to the crop—a giant sac that can expand to accommodate everything that the insect sends its way (which is just about anything).

A cockroach's biggest secret is the teeth in its stomach. Technically, these are chitinous denticles in the cockroach's gizzard that work the food over for a while. Released nutrients are absorbed by fingerlike projections. What-ever's left flows into the colon, where food by-products mix with metabolic wastes absorbed and released into the gut by the roach's spaghetti-like Malpighian tubules, which are excretory organs common to all insects. In the last and most familiar step of this elaborate process, water is removed from the gut contents by rectal pads, producing dry, pelleted feces, that depart through the anus.

FAT BODY:

Another secret of cockroach success is the fat body, a lumpy white organ that fills almost all available space in the abdomen. This mass of cells serves several functions: it's the insect's warehouse (in which protein, glycogen, and fat are stashed in anticipation of lean times); a factory (where amino acids and vitamins are made); and a recycling plant (for transforming a waste prod-uct, uric acid, for reuse as an energy source).

REPRODUCTION:

The interior ductwork of male and female cockroaches is refreshingly similar to our own—leading respectively to sperm-producing testes and egg-bearing ovaries. Fertilization is internal, but that's where similarities to the human reproductive system end (more on that in chapter five).

Recent research has established that a cockroach breaks wind every 15 minutes. It also continues to release methane for 18 hours after death. Insect flatulence is said to account for 20 percent of all methane emissions on earth, placing termites and cockroaches among the biggest contributors to global warming. Writing about this phenomenon, University of Illinois entomologist May Berenbaum downplayed its harmful consequences. "Tiny gas bubbles are even visible in Dominican amber, clinging to the abdomens of termites, cockroaches, millipedes and other gassy arthropods," she observed. "The process has been silent for millions of years, but it hasn't proved deadly yet."

SENSATION:

A cockroach is truly the beast with two brains. It has two pairs of large nerve ganglia in its head, as well as a single nerve ganglion in its tail. These two sensory centers are connected, ultimately, by giant fibers. The components of a neural info highway, these giant fibers carry impulses ten times faster than ordinary nerves—they travel the length of a roach's nerve cord in around .003 seconds. This arrangement gives cockroaches the ability to turn sensory input into action in record time. Experiments have shown that a warning message from the cerci could be translated into rapid leg movements in .045 seconds—literally faster than a blink of the human eye. The faintest breeze that precedes a rolled-up newspaper on the downswing is all it takes to give roaches a running head start.

A whimsical advertisement from a Victorian watchmaker.

How Smart Are They?

Cockroaches can learn to navigate mazes, successfully working their way around assorted bends, switchbacks, and dead ends after only five or six trials. This ability was first reported by the American biologist C. H. Turner, who found that his lab-reared cockroaches could memorize complicated routes in successive trials over the course of a day. However, they soon forgot their lessons, requiring Turner to retrain his prize pupils at the start of each session. Subsequent researchers have had better luck with learning and memory in other cockroach species.

Another researcher suspended roaches over a saline solution; whenever one of these insects dipped its leg in this solution, it completed an electrical circuit and received a mild shock. After about thirty minutes, they learned to keep their legs in a raised position, thus avoiding getting zapped. Decapitated specimens reacted to shocks in much the same manner—a testimonial to the insect with a brain in its behind.

Results from both experiments put cockroaches on the shortlist of brainy invertebrates, not too far below the octopus, which can be trained in little time to associate certain complex symbols with offers of food or threats of electric shock. Measuring the mental abilities of roaches is no less challenging or controversial than determining the intellectual capacity of our own kind. It's safest to say simply that they are as smart as they need to be. Let's leave it at that.

Life span

Cockroaches can survive for a single season or several years, depending on the species. Even within a species, there's considerable variation in the number of days that these animals can survive. Rates of maturity depend on air temperatures, moisture levels, food resources, and other environmental conditions. In many species, eggs hatch in early summer, and the young take a hiatus, maturing the following spring.

Longevity is also influenced by environmental conditions, and varies within species:

Oriental cockroach	316-533 days
American cockroach	100-500 days
German cockroach	90-200 days
Brownbanded cockroach	115-136 days
Smokeybrown cockroach	191-586 days
Giant cockroach	300-600 days

Aggressive tendencies

Some cockroach species aren't that much nicer to each other than we are to them. Observations of laboratory cultures of American cockroaches have shown adult males to be extremely aggressive, often biting and kicking other adult males and, occasionally, adult females. Female-female encounters are nearly nonexistent in this species. However, among female German cockroaches, levels of aggression are equal to or slightly higher than those of males.

Aggressive encounters between most species studied begin with two cockroaches meeting head-to-head for a brief bout of antennae-fencing (see p. 66). After this initial contact one or both of the opponents may adopt threat postures. Included in the cockroaches' repertoire of aggressive poses is "stilt-walking" (in which an aggressive individual straightens its legs, raising its rigid body high off the ground) and "body-jerking" (pretty much like it sounds).

Occasionally, these displays lead to a mutual truce. However, they are most often followed by one-sided aggressive acts, either charging, biting with the mandibles, or kicking with fore- or hindlegs. These usually prompt the recipient to retreat— a move that most often terminates any further hostile acts.

Tussles over turf

Intense battles can erupt should both cockroaches choose to stand their ground. During these protracted bouts, the combatants will stalk in circles, stilt-walking and swapping bites or kicks for up to three minutes. "These movements were very fast and often occupied the entire floor of the cage," observed William Bell, in his paper analyzing aggressive encounters between American cockroaches. "Some idea of the fierceness of the battles can be gathered from the fact that sometimes, though rarely, legs were torn off." Of the nearly 600 aggressive encounters witnessed by Bell, only seventeeen reached this level of intensity. None of these resulted in the death of either combatant.

Bell was unable to predict the outcome of any of his cockroaches' aggressive encounters. He did, however, learn that aggressive tendencies of American cockroaches are closely related to sexual activity or territorial defense, which in the case of male cockroaches are clearly linked. Other researchers

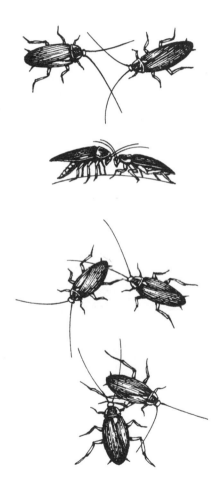

The stages of a cockroach fight (top to bottom): antenna fencing, stiltwalking, kicking, biting.

have determined that the intensity of aggression among cockroaches is directly correlated to population density.

It's been suggested that the high incidence of aggression among gravid adult German cockroaches arises from these animals' instinctive urge to safeguard their egg cases and nymphs from cannibalism (see p. 102). Studies of other cockroach species have revealed additional reasons for combative behavior: by gaining control over territories in which food and water are clumped, the males of various species can starve their competition, thus claiming for themselves any receptive females that come to these well-defended oases.

Where Are (or Aren't) They Found?

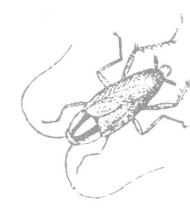

LIKE THE QUANTITY OF HAMBURGERS and cheeseburgers McDonald's advertises having sold, the number of recognized roach species keeps growing steadily—if not by the millions. New blattarians are added to the list at the rate of around forty per year, bringing the current total to around 3,500 species. One world-renowned authority on our friends the Blattaria, Dr. Louis M. Roth of Harvard University's Museum of Comparative Zoology, has single-handedly identified more than three hundred species, and all of those within the past ten years. Many entomologists believe that the number of recognized cockroach species may eventually reach five or six thousand.

The vast majority of the species that have been discovered and indentified to date are confined to a broad band of land some thirty degrees north to about thirty degrees south of the equator. Within this warm, photosynthetically supportive zone are the world's most productive and botanically diverse forests: from a single twenty-five-acre plot on the island of Borneo, for example, scientists have tagged 780 species of trees—nearly sixteen times the number from all of Washington, which proudly proclaims itself the Evergreen State. In these lush equatorial regions, food is plentiful for plant-eating animals of all shapes and sizes, inlcuding cockroaches. Temperatures are nurturing, usually over seventy-five degrees, and moisture is high throughout most of the year.

A blattarian bounty

What constitutes an abundance of cockroaches? The small South American country of French Guiana boasts a total of at least 118 species while the tropical forests of Costa Rica may have as many as 150 species, according to Frank Fisk's *Annotated Checklist of Costa Rican Cockroaches*. C. F. A. Bruijning of the Netherlands recorded 376 species from the Malayan subregion—Java, Sumatra, Borneo, the Malay Peninsula as far as the isthmus of Kra, and the many smaller adjacent islands. In his opinion, the species on his list represented "only a part of the vast number" of natives to this roach-rich region.

Data on the abundance of African cockroach species are not readily obtained. However, it is safe to assume that this continent's fauna is comparable to French Guiana's or Costa Rica's. Why else would the West African Republic of Equatorial Guinea issue a four-color postage stamp commemorating the American cockroach in all of its glory?

With each mile to the north or south of the equator, the conditions become progressively less favorable for roaches. South Africa has around 125 indigenous species, and in all of Australia there are 175 such natives. With the exception of major metropolitan centers such as Montreal or Toronto, all of Canada is fairly roach-free. Centuries of trade with Asia, Africa, and the Middle East have brought many non-native, pestiferous cockroaches to northern Europe. However, even with these intruders, there are still less than seventy-five species in the entire continent—half the number of roaches in the tiny nation of Costa Rica.

America's least wanted

The cockroach fauna of the United States contains members of all five families, and represents a total of thirty-one genera and sixty-nine species. More than two-thirds of these species have been introduced from outside.

There are no indigenous cockroaches in Idaho, Montana, or Wyoming. Forests in Washington and Oregon harbor two natives—the western wood cockroach, *Parcoblatta americana*, and a newly identified species of brown-hooded wood cockroach, *Cryptocercus clevelandi*.

Woodland habitats in the northeastern United States have as many as four species of *Parcoblatta*—the fulvous (*P. fulvescens*), Uhler's (*P. uhleriana*), Virginia (*P. virginica*), and Pennsylvania (*P. pensylvanica*) wood cockroaches. Habitats in the southeastern states also contain these insects, plus the broad (*P. lata*), banded (*P. zebra*), Boll's (*P. bolliana*), desert (*P. desertae*), southern (*P. divisa*), and Caudell's (*P. caudelli*) wood cockroaches.

Fifteen more natives (not all of them wood cockroaches) are found in Texas, making this the second most cockroach-rich state in the Union. The prize for the greatest diversity of native cockroaches goes to Florida, with a total of twenty-seven native species. Included in this tally are a couple of real collector's items—the glossy claret-brown broad Keys cockroach (*Hemiblabera tenebricosa*), the translucent-yellow Florida beetle cockroach (*Plectoptera poeyi*), and the handsomely patterned small hairy cockroach (*Holocompsa nitidula*) from Key West.

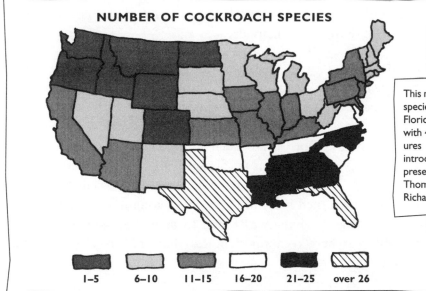

NUMBER OF COCKROACH SPECIES

This map shows the number of different species found in each state of the union. Florida and Texas are by far the winners, with 41 and 36 species respectively (figures include both native and exotic introduced species). Adapted from data presented in *Pest Control* magazine by Thomas Atkinson, Philip Koehler, and Richard Patterson.

1–5   6–10   11–15   16–20   21–25   over 26

Peridomestic cockroaches

Of the world's 3,500 cockroach species, only fifty are considered domestic pests. More than half of these are considered peridomestic nuisances—that is, they prefer to live around but not actually *in* houses and other artificial structures.

One of these peridomestics is the Pennsylvania wood cockroach (*Parcoblatta pensylvanica*), a native of eastern, southern, and midwestern North America. Ordinarily this species dwells in oak, chestnut, and pine forests (hence the "wood" in its common name). However, it becomes a seasonal inhabitant of homes, especially during winter months when it is accidentally carried indoors with armfuls of firewood. One female Pennsylvania wood cockroach can produce up to thirty egg capsules, each containing thirty-two to thirty-six eggs. Thus, it is not uncommon for an indoor epidemic to be triggered by just one of these uninvited visitors.

A gang of five

In the United States, a total of five domestic roach species have garnered the lion's share of attention—the pesky American, German, smokeybrown, Oriental, and brownbanded. None of these are native to this country; they were unintentionally introduced many decades ago.

AMERICAN

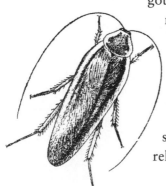

The American cockroach needs no introduction—and besides, it already got one on page 6. Quite possibly the most widely distributed of the cockroach pests, it is also the most frequently encountered resident of urban sewer systems. Once established in these dank environments, these roaches have little reason to disperse. In one classic study from the early 1950s, sixty-five hundred radioactively labeled American cockroaches were turned loose in Phoenix, Arizona underground. Sixty days later, only one specimen was found to have migrated out of the sewer system. Researchers recaptured it less than 125 feet from its original release site.

GERMAN

A member of the Blatellidae, the German cockroach (*Blattella germanica*) is also known as the croton bug, after New York's Croton Valley Aqueduct, where these insects greatly extended their range in the 1800s via the conduits and pipes. Adults are light brown, with two dark, parallel stripes on their backs. Both sexes have wings, but seldom fly. Young are wingless, with single stripes and much darker bodies than their parents.

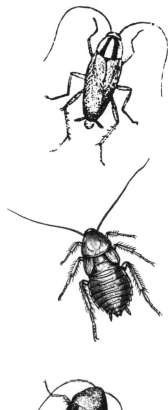

SMOKEYBROWN

Like its close relative the American cockroach, the smokeybrown (*Periplaneta fuliginosa*) is particularly difficult to control because in many parts of the world new indoor populations are consistently recruited from the out-of-doors. A study conducted at a suburban home in southeast Texas showed that gravid females routinely came in from the cold during October and November, presumably in an attempt to find warm winter harborages for their egg cases. Nearly 26,000 smokeybrown females and their mates were captured during this sixteen-month investigation of the 133-square-foot homesite.

ORIENTAL

Only male Oriental cockroaches (*Blatta orientalis*) have functional wings; females have short stubs and are permanently grounded. Despite this apparent handicap, this Asian species has extended its range to include every state and province of North America, as well as all of Europe.

BROWNBANDED

Of African ancestry, the brownbanded cockroach (*Supella longipalpa*) made its first U.S. appearance in Florida nearly a hundred years ago. But it wasn't until after World War II, when soldiers returned to the States carrying roach-infested duffel bags, that these household pests gained a firm foothold. By 1967, brownbandeds were reported in forty-seven of the forty-eight contiguous states. These insects are also known as TV roaches because of their fondness for warm spots, which often include the heat-generating innards of electric appliances. Body fluids from dead brownbandeds can create short circuits in computers and electronic equipment.

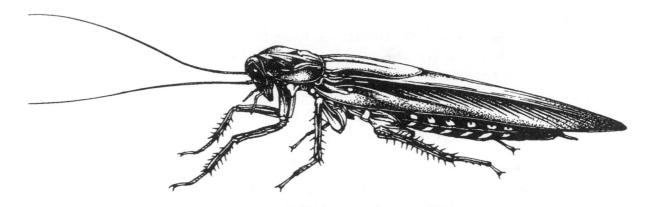

The winner: *Megaloblatta blaberoides*

The world's biggest cockroach

Some will tell you that "the roaches around here are bigger than the rats." But that is a gross exaggeration. No cockroach has ever been measured over five inches—about half the length of a healthy Norway rat (*Rattus norvegicus* in case you're interested).

So what constitutes a giant cockroach? If one is to believe the 1995 *Guinness Book of World Records*, the largest specimen on record is a preserved female *Megaloblatta longipennis* from Colombia, now in the collection of Akira Yokokura of Yamagata, Japan. This magnificent creature measures 3.81 inches (97 mm) from head to stern. This is certainly no rat-sized critter but it's not all that much smaller than a common field mouse or a shrew.

Of course one shouldn't always believe the *Guinness* editors, since they apparently overlooked "The Largest Cockroach," a brief paper by Ashley B. Gurney, published in volume 61 of the *Proceedings of the Entomological Society of Washington* in 1959. This paper provides documented proof of a 3.94-inch (100 mm) adult female *Megaloblatta blaberoides*, a full eighth of an inch longer than Yokokura's prize specimen, and the real ascendant to the whopperdome throne. The wings of this giant, a common inhabitant of the Central and South American jungle, are more than seven inches from tip to tip. What does an insect this size eat? Obviously, just about anything it wants.

Despite these two, irrefutable records, Australian experts continue to brag that their native giant burrowing cockroach (*Macropanesthia rhinoceros*) of North Queensland is the biggest in the world. Technically speaking, they're correct: while *Macropanesthia* males and females are a full three-quarters of an inch shorter than either species of *Megaloblatta*, they are by far the heftiest roaches on record, weighing nearly one and a quarter ounces. That's about the same weight as a (AA) battery—or three dozen adult American cockroaches.

As their common name implies, giant burrowing cockroaches are diggers, constructing underground chambers one to three feet down, where the temperature is a comfortable twenty degrees centigrade year round. Here they live on fallen twigs and dry leaves, gathered from the litter that surrounds their deep dens. Their digs are often shared with centipedes, beetles, silverfish, other roach species, and, occasionally, large frogs.

Hard, stout spines on its front legs help the giant burrowing cockroach excavate earth like a mole, flicking soil to the rear as it digs. While moving through an established burrow, these spines can be conveniently folded out of the way. The cockroach's shovel-shaped pronotum also serves as a digging tool.

Adult females produce up to thirty young, which are born alive around November of each year. These stay with their mothers for about nine months, then set out on their own, each building a burrow that they'll inhabit for the next three or four years. Adult males frequently cohabit with nearly grown females, apparently waiting for them to mature. After mating with these promising young things, the males leave the burrows and look for the chambers of other virgin females.

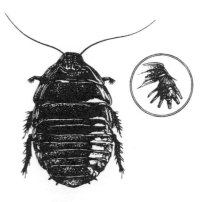

The modified foot of *Macropanesthia* is for digging.

The real record holders

In July 1986, when the Bizzy Bees Pest Control Company of Dallas, Texas, offered a thousand dollar prize for their state's largest roach, they received an overwhelming response. Fourteen finalists, nine of them dead, were selected from a field of 205 entrants. In accordance with contest rules, all of the entrants were American cockroaches. No giant foreign species were allowed.

The winner was an impressive 1.92-incher entered by Pat Camden, a draftswoman for Southwest Bell in Dallas. Camden and two other co-workers used a spray insecticide to subdue the roach as it brazenly marched down a corridor at her work. "It was kind of sad when it died," she confessed in an issue of *People* magazine. "I guess I ought to kiss it, huh?"

Fame was short-lived for Camden and her friends, who were soon overshadowed by a 2.42-inch insect from Florida—the wood cockroach winner of the Orlando Sentinel's competition, held the following month. New York, Philadelphia, and other U.S. cities were fast to follow suit. By 1989, after Michael Bohdan, co-creator of the Bizzy Bees' event, moved his business to Plano, Texas, the search had become international in scope, with entrants from South Africa and Bermuda.

Today Bohdan no longer focuses solely on roaches of unusual size. At his slightly eccentric Cockroach Hall of Fame, he hosts an annual "Best-Dressed Cockroach" contest, attended by such luminaries as "Liberoachi" and "Marilyn Monroach." "Be creative and please, for the sake of the fainthearted, use dead roaches," contest fliers proclaim.

"Why can't a city big enough to bid for the Olympics scare up a world-class cockroach?" groused columnist Jim Auchmurty in a 1990 issue of the *Atlanta Journal*. Auchmurty cited the city's four previous attempts at scoring the biggest roach, adding that repeated failures to capture the international bigness cup were "not for lack of specimens, as anyone with a damp basement can attest."

At home with ants

At the opposite end of the scale from *Macropanesthia* is the 3/32-inch long *Attaphila fungicola*—a roach that's roughly the size and shape of a peppercorn sliced in half. One of several dozen kinds of blattarians that live in close association with social insects (ants, termites, wasps, and bees), this animal lives year-round in the underground nests of tropical leaf-cutting ants. It feeds on the fungi that the ants actively cultivate for their own nutritional purposes, and, in return, may provide limited janitorial services for its nestmates.

Both male and female *A. fungicola* are known to hitch rides on the backs or heads of the large soldier ants in the nests. The legs of the females have highly developed sticky pads that enable them to cling to the winged forms of the leaf-cutters during their seasonal swarms. In this manner, the reproductively active female roaches are taken on a wild ride to other parts of the tropical forest, where they quickly make themselves comfortable in their carrier's new nest. Males quite possibly find new homes by following scent trails left by other ants traveling on foot.

Canopy cockroaches

Few people would think of looking for cockroaches high above their heads. But that's exactly where biologist Donald Perry found them, in the tree-tops of Costa Rica's La Selva wilderness. Here, according to Perry's firsthand account of *Life Above the Forest Floor*, "large, ground-zone type cockroaches have given way to smaller, aerial varieties endowed with unexpected beauty." When Perry began his research in the early seventies, the study of forest canopy life was solely for brave pioneers. To gain access to his study sites, which were sometimes as high as 200 feet in the air, the enterprising biologist tied nylon climbing ropes to arrows, then, with a crossbow, shot them into the overstory.

More down-to-earth methods for studying canopy cockroaches and other arboreal insects have been employed by biologists in Panama's lowland forests. Terry Erwin of the Smithsonian Institution has modified a commercial Dyna-fog fogging machine for controlling mosquitoes, hoisting it into the canopy, and blasting everything with insecticide for a full minute. The shower of dead insects from this spraying rains down onto sheets spread beneath the trees. Individual specimens can then be gathered, sorted into family groups, and identified.

Ohio State University entomologist Frank Fiske has looked at specimens similarly acquired to determine the relative amounts of canopy cockroaches at various times of the year. In the late dry season, he counted 218 adult and immature roaches from sixteen different species. A few months later, in the early wet season, he counted over four times this number—a total of 991 individuals—from seventeen identified species. His figures showed that roaches in portions of Panama's forest canopy lay low during the summer's dry seasons, then come alive after the first hard rains in the fall.

Cockroaches in caves

Caves in tropical forests afford cockroaches near-ideal conditions, encouraging them to thrive and multiply. Humidity is consistently high, so drinking water is always available in the form of condensation on the cave walls. There is usually plenty of food—numerous fungi, plus the droppings and dead bodies of bats and other cave cohabitants.

In 1980, when David Brody, a technician with the Entomology Department at the American Museum of Natural History, landed an assignment to collect roaches for the motion picture *Creepshow*, he headed to one of these dark recesses on the Caribbean island of Trinidad.

Accompanied by his associate, Raymond A. Mendez, Brody stumbled upon a true "glory hole"—a section of a 200-foot cave, deeply encrusted with bat guano upon which thousands of giant cockroaches fed. The blattarians were so thick, according to Brody, that the ground actually undulated with their movements.

Not wishing to remain in this dank domain any longer than necessary, Brody and Mendez wasted little time in filling their plastic bags with cave cockroaches. Eventually they extracted two thousand specimens, shipping these back to the U.S. in egg boxes strengthened with thin plywood frames. The fruits of their labors can be seen in one of *Creepshow*'s final scenes: a grotesque tableau with thousands of roaches flowing from a fissure in film star E. G. Marshall's forehead.

The giant cockroach is one of a handful of species that commonly frequent caves but are in no way limited to these habitats. Many other species have evolved into complete creatures of the dark. *Nocticola caeca* is one of these, an anemic-looking insect from caves on the island of Luzon in the Philippines. Like many cavernicolous invertebrates, this species is born without eyes and receives information about its surroundings from other senses, primarily touch and smell.

Cockroaches in cities

One widely accepted rule of thumb holds that in a big city there are at least ten cockroaches for every human. If one can accept this supposition (and, really, how could anyone ever prove such a thing?), there are some sixteen million roaches in Houston and at least fifty-seven million more in a city the size of New York.

A more realistic measure of blattarian populations in major U.S. cities is the amount of money spent on their control. For this information, we can all be grateful to the makers of Combat cockroach-control products, who compiled these figures for 1994:

> Players of the computer game **Bad Mojo** are transformed into cockroaches and given a true insider's view of a seedy San Francisco hotel and bar.

CITY	COCKROACH INSECTICIDE SALES (IN MILLIONS)
Los Angeles	$15
New York	$10
Houston	$7
Miami	$6.3
Dallas	$6.1
San Antonio	$5.7
Baltimore	$4.7
New Orleans	$4.6
Tampa/St. Petersburg	$4.5
Birmingham	$3.5
Orlando	$3.0
Atlanta	$2.9
San Diego	$2.5
San Francisco	$2.3
Philadelphia	$2.2
Phoenix	$1.9
Chicago	$1.8
Richmond	$1.8
Raleigh	$1.7
Jacksonville	$1.7
Charlotte	$1.3
Sacramento	$1.1
Roanoke	$.83

The art of estimating

It's extremely difficult to guess the size of an indoor roach infestation, particularly when new recruits are constantly being enlisted from other parts of the building or brought in from the outside. To arrive at a scientifically acceptable number for such populations, researchers often use what is called the "mark-recapture" technique: cockroaches are caught and color-coded with dabs of paint or labeled with radioactive isotopes. Then they are released and allowed to return to their preferred dwelling places. When a predetermined amount of time has elapsed, a second sweep is staged, with researchers nabbing as many of the insects as they are able. After counting the number of roaches from the first batch that have been recaptured, a statistical formula can be applied to determine the approximate size of the total cockroach population.

More often than not, researchers and pest-control operators use their best professional judgment, in many instances based on several decades of observation in the field. Such informal methods were employed by Dr. Austin Frishman (a.k.a "Dr. Cockroach") as part of a contest sponsored by Combat Labs. In 1994 and again in 1995, he and his comrades at Combat sifted through hundreds of entries from thirty states, selecting what they deemed the nation's six worst infested homes. The winners—in New York City, Philadelphia, Dallas, Mobile, Tulsa, and Jacksonville—received free treatments, overseen by Dr. Cockroach himself.

The cockroach population at one of the winning households was so pervasive that family members nearly abandoned their home, sleeping in their Suburban van and dining almost exclusively at fast-food restaurants. According to *The Dallas Morning News*, Frishman gave the house a rating of two and a half on a scale of one to five (five being the worst), estimating that it hosted more than ten thousand roaches. At another site, residents slept with the lights on and shook cockroach nymphs from their shower curtain every morning. Frishman gave this one-story house a three, appraising the indoor population at between sixty and a hundred thousand roaches. A five, he said, is "when you walk into a room and there are thousands of exposed cockroaches and several hundred thousand hidden from view"—something he's seen only in the most poorly maintained house.

Cockroaches were the "bird dogs" that put FBI investigators on the trail of Patty Hearst's kidnappers in 1974. It all began when the landlord of a downtown San Francisco apartment building responded to a tenant's complaint of cockroaches "swarming down the walls" from the suite above. Entering the abandoned flat, the landlord discovered what was later identified as a former hideout used by the Symbionese Liberation Army. This discovery, initiated by cockroaches, proved to be the first solid lead in the famous case according to *Patty/Tania* by Jerry Belcher and Don West (Pyramid Books, 1975).

Cockroaches in Congress

In 1982, entomologists Phil Koehler and Richard Patterson of the U.S. Department of Agriculture's Gainesville lab were summoned to Washington, D.C. Their mission: to curb a German cockroach infestation at the House of Representatives. For months, a particularly hard-to-kill strain had been colonizing coffee machines and lounging in lunch rooms at the House and adjacent office buildings. Washington's resident pest-control experts were stymied. What could Koehler and Patterson recommend?

The two entomologists collected a few hundred specimens from the Congressional chambers, then flew them back to Gainesville for pesticide testing. Here it was determined that this strain had become invulnerable to most pesticides, as a result of years of continuous spraying. "We needed to find out what chemicals they were resistant to, before we could make recommendations about what next they should spray," Phil Koehler remembered.

Today, the descendants of these roaches (roughly fifty-six generations, as of this writing) are maintained by the U.S.D.A. and a few commercial laboratories, including the S.C. Johnson Corporation in Racine, Wisconsin. The name of this strain is HRDC—an acronym that simply stands for House of Representatives, District of Columbia.

"We use HRDC as a reality check because it's resistant against so many different chemicals," Johnson's Director of Insect Development, Keith Kennedy, told readers of *Scientific American* in 1994. "It seems to be a super-tough cockroach and it's only fitting that it came from Washington."

Whether Koehler and Patterson ever resolved the Congressional roach problem is unclear. But even with the right insecticide sprays, it's doubtful that anyone could drive all the verminous creatures from either the House or the Senate.

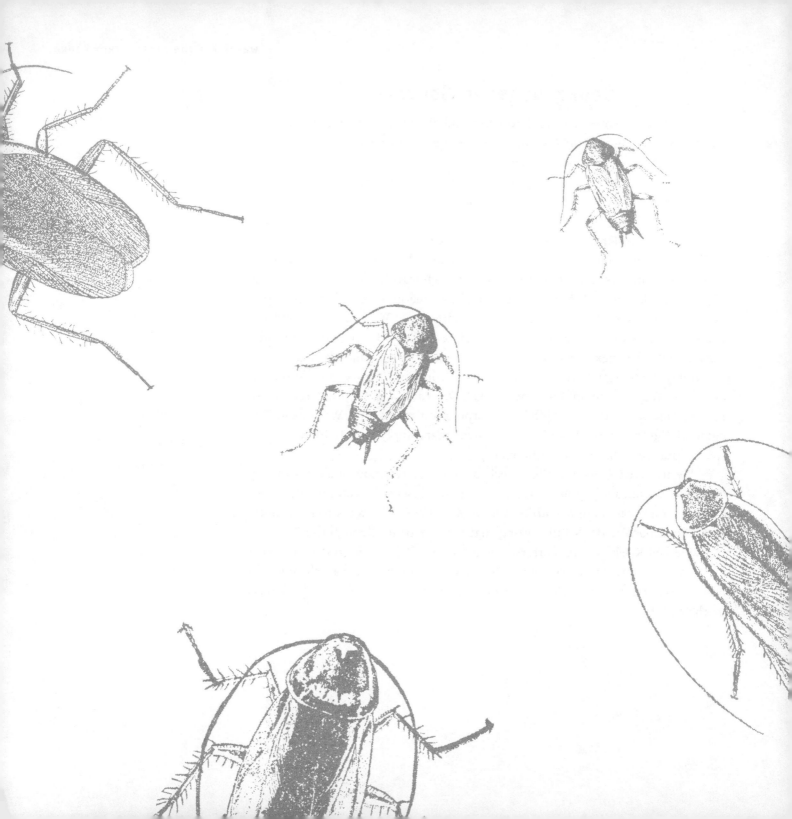

A 340-Million-Year History

GEOLOGICAL EVIDENCE, largely in the form of fossilized wings, indicates that cockroaches have been around for at least 340 million years. At that point in our planet's five-billion-year history (during a period in the Paleozoic called the Early Carboniferous), the land masses we recognize today as North America, South America, Africa, Australia, Asia, Eurasia, and Antarctica were all part of a single large continent. Much of this continent's interior was marshy and supported the great forests from which the earth's coal deposits were formed.

Carboniferous forests were vastly different back then: instead of tall trees (which had not yet appeared on the planet), timber stands were largely composed of giant fernlike plants and club mosses, some sixty to a hundred feet tall. Beneath this strange green canopy grew a few puny upstarts—the ancestors of our present-day cycads and ginkos, and the diminutive forebears of our modern conifers. There were no colorful flowers or tasty tropical fruits, for these, too, had yet to evolve. Seeds, nuts, nectar, and pollen would not appear for at least another 200 million years.

The first primates, from which *Homo sapiens* would gradually evolve, would not appear on our planet for another 300 million years. Dinosaurs were also a long way off—some 150 to 180 million years in the future. Nonetheless, there *were* roaches, perhaps as many as 600 species, thriving and multiplying.

The fossil remains of these insects are so plentiful that paleontologists have nicknamed the Carboniferous "the Age of Cockroaches." Over 1,900 specimens have been removed from a single site in Coventry, France. Finds such as these suggest that during the Carboniferous, cockroaches may have accounted for as much as forty percent of all insect life. (Some scientists maintain, however, that the large, heavily armored bodies of cockroaches, plus their penchant for moist, swampy areas (within which fossils are more frequently formed), made these animals better suited for preservation in stone.)

Cockroach ancestors

Ancient blattarians looked remarkably like their descendants, the cockroaches of today. They had long antennae, stocky but flattened bodies, and heavily veined, somewhat paddle-shaped wings. Unlike modern roaches, these insects were equipped with long tubes for laying eggs, one at a time, either on the ground or under the bark of tree ferns and other large plants. This tube (called an ovipositor) was often as long as the roach's body. The wings of ancient cockroaches were also folded a bit differently, and the abdomens of the immature forms were more slender and supple. However, those are the only real differences between Carboniferous roaches and their modern-day equivalents.

Live and decaying plant matter was plentiful in the Age of Cockroaches, so few blattids ever had to go without meals. Insect predators weren't even a concept yet; murderous parasitoid wasps (see page 108) had yet to evolve. Any roach that could avoid spiders, centipedes, and a few shore-dwelling fish had it made.

As Earth's single landform slowly divided over millions of years, and seven continents were created through the phenomenon known as continental drift, ancestral cockroach stocks became geographically isolated. New species evolved while others became extinct. Entirely new insect families—including the beetles, bees, termites, and ants—became prominent figures in the fossil record. It was around this time—roughly 100 million years ago—that the founders of the five modern roach families are thought to have emerged.

As the relationships between plants and these comparative latecomers became more specialized and complex, encouraging the number of insect

> "According to tradition, our ancestors were descended in early Cretaceous times from certain kind-hearted cockroaches that lived in logs and fed on rotten wood and mud."
>
> —the king of the *Termes bellicosus* termites, as translated by William Morton Wheeler in *Foibles of Insects and Men* (1928)

species to increase by leaps and bounds, the roaches' relationship with the world remained unchanged. The forebears of modern birds, reptiles, amphibians, and small mammals leaped onto the scene by the thousands, applying progressively more pressure on the population. Today cockroaches represent less than 1 percent all living insects—down quite a few points from the days of yore.

"Once a cockroach always a cockroach is their motto," wrote William Beebe, the New York Zoological Society's most eloquent biologist, and author of *Jungle Days*. "They are found everywhere, contented with a safe, middle course of life, seldom aspiring to size or bright colors, never attacking or even defending themselves, or putting on side in their life-histories."

Other ancient insects

A few fossils of insects as old or even older than roaches have been unearthed. The earliest of these, *Rhyniella praecursor*, is from the Devonian period, which began roughly forty million years before the Carboniferous. This fossil possesses all the major characteristics of contemporary insects: three pairs of legs, segmented antennae, and a body divided into three parts.

Other fossils from this period are of wingless animals resembling modern-day springtails and silverfish. These primitive creatures almost certainly lived in (and fed on) moist, rotting plant debris at the edges of shallow Devonian lagoons. Dragonflies are also as ancient. Judging from their fossil remains, some of these insects were enormous, with wingspans of two feet—a full sixteen inches longer than those of any living species.

A living fossil

The brown-hooded wood cockroach, *Cryptocercus punctulatus*, is most closely related to the ancient stock from which modern cockroaches evolved. This insect's mahogany-colored body is about an inch long (23–29 mm), and is speckled on its back and sides with numerous small pits. The pits are probably associated with the production of pheromones—aromatic compounds released by roaches and other insects to attract or repel their own kind. Both males and females are wingless, with short, stocky legs and nearly nonexistent antennae. Devoid of any and all frills, these primitive specimens more closely resemble termites than blattids.

C. punctulatus also acts like a termite, supporting the notion that the two insect orders shared a common ancestor, many millions of years in the past. Unlike most other insects, adult brown-hooded wood cockroaches bond for life. Together they chew into rotting wood a series of small galleries with larger rearing chambers, which serves as their sole family residence for the rest of their lives. Both sexes tend to their young, which are usually born a year after these insects pair off. Broods are quite small, seldom exceeding three or four young.

Shortly after their brood hatches, the parents perform a very termitelike act—they pass special cellulose-digesting protozoans to their young, in this case through special intestinal fluids, upon which the young feed. Only with these protozoans can their offspring obtain nutrients from their diet of decaying wood.

The family stays together for three to four years, during which time the youngsters mature, undergoing successive color changes from white to ivory, to gold, to reddish brown, and, finally, dark brown or black. The parents continue to feed and tend to their young throughout this gradual transformation. Only at the end of their parents' six-year life span are the offspring required to fend for themselves.

Cockroaches in human history

While the majority of cockroaches are content with their niches in the wild, some have seized the opportunity to move indoors. This bold step may have been taken two million years ago, when our distant ancestor, *Homo erectus* (also known as the Java or Peking Man), sought shelter in the caves of tropical Asia and Africa. By building lean-tos and other crude shelters on these continents, early humans quite literally opened their doors to roaches. These structures offered all the amenities of cave habitats and then some; once people learned to fill their larders in anticipation of seasonal shortages, they inadvertently guaranteed a lifetime supply of nutritious food for any freeloaders small enough to squeeze through the cracks.

"It will surprise the reader to hear that the following natural-history objects have been found hidden in banana bunches in the city of Minneapolis: a young snake (Boa constrictor), 14 inches long; a large crab, four inches across the back; two species of scorpions; several large thousand-legs (Scolopendra); some large and black cockroaches; several large and hairy bird-spiders, usually called tarantula, and many other smaller insects and other beings."

—*Grasshoppers, Locusts, Crickets, Cockroaches, Etc. of Minnesota,* Agricultural Experiment Station, Entomological Division, University of Michigan (Bulletin No.56)

As our predecessors founded new village sites across Africa, Asia, and the Middle East, they carried cockroaches with them. Such hospitality enabled these species to colonize otherwise unlivable habitats. Today, roaches, like people, can be found almost everywhere—from mountaintop shelters in Norway and Switzerland to the steam tunnels of Fairbanks, Alaska. In the 1920s, a sizable population of American cockroaches even turned up in a coal mine near Glamorganshire, England, thriving at a depth of 2,166 feet. It's possible that these ubiquitous insects have already traveled in outer space.

An early cockroach chronicler

"Most men talk much of the Blatta, but few or none able to describe what the Blattae, properly so called, are," begins Chapter XVIII of *The Theater of Insects: or, Lesser living Creatures*, by Thomas Mouffet, first printed in 1658. A Doctor in Physick, whose fascination with spiders and insects is immortalized in "Little Miss Muffet" (one of several satirical nursery rhymes attributed to the legendary Mother Goose), Mouffet laid a number of rumors to rest: the Blatta are not "the worms growing in the ears," as the Roman scribe Pliny the Younger maintained, nor are they "a little worm eating cloathes or books," as the poet Horace claimed.

"Now the Blatta is an insect flying in the night," Mouffet told the scholars of his day, "like to a Beetle-bug, but wanteth the sheath wings." He then went on to describe the three living forms of blattids. Because the word cockroach had not yet been incorporated in the English language, he called all three of these animals "moths." Nonetheless, his characterization of the blattids as "fliers of the light, nasty, cruel, rough, theeving, living of nocturnal depredations after an infamous manner," as well as his description of their habitat ("about Privies or Jakes houses, ditches and steamy soggy places"), makes it easy to recognize these insects for what they truly were. Mouffet concluded his chapter on roaches with the observation that "God hath endued them with sundry vertues," somehow failing to establish what these might be.

The "soft Moth," according to Mouffet, "hath a small head, whereout comes two long cornicles, every waies movable," a forked tail that resembled "a pair of Barber cizzers," and "wings of the flame color." The Mill- or bake-

Woodcuts of roaches from Mouffet's *The Theater of Insects*

37

house Moth "is longer, thicker, and of a more shining black colour than the ordinary soft Moth, with a little forked mouth placed as it were under its belly." The "unsavory or stinking Moth" is wingless, its body a glistening black, and "doth not only annoy those that stand near it, but offends all the place thereabouts with its filthy favor."

Known today as the Oriental cockroach, the "stinking Moth" probably reached Europe by way of the Silk Road, traveling at a camel's pace from the markets and bazaars of central Asia and the Middle East. Predisposed to a much warmer climate, this animal sought shelter indoors, finding cozy niches in wine cellars, warehouses, and even, according to Mouffet, the uppermost parts of a church in Merry Olde England. In Mouffet's day, the Oriental cockroach was also called a "black beetle" or "black clock," the latter name referring to the timepiecelike precision with which this insect appeared at sunset. Many people believed that if one of these beasts flew into a parlor or bumped into you at dusk, severe illness or perhaps death would follow.

A second insect immigrant, the German cockroach, may have hitchhiked across Europe in breadbaskets carried by soldiers returning from Prussia after the Seven Year's War (1756-1762). To this day, Russians call this particular species *Proussaki* (the Prussian cockroach) and distinguish it from other species, which they commonly call *tarakan*. "There are loyal Russians who will tell you solemnly that there never used to be cockroaches in Moscow, that the little beasts were first seen at the Kievsky station in 1955, when the first trainload of African students arrived and opened their suitcases for inspection," reports Martin Walker, an American journalist in Moscow during the heyday of the Soviet Socialist Republic. "This is rubbish," Walker exclaimed, citing the old town of Tarakanov, due north of Moscow, as firm proof of the roach's long-standing membership in the Communist Party. "The cockroach is as Russian as borscht."

Cockroaches at sea

Around the same time that Mouffet's book went to press, seafarers from England, Spain, Portugal, and other nations were busy ferrying roaches from one exotic port of call to the next.

This lithograph from Victorian England shows roaches
in their prime, cavorting by the sea.

With plenty of food and water, ample cracks and crevices to fill, and few
natural predators to avoid, the insects aboard these frail wooden vessels had
it made. In 1587, when Sir Francis Drake captured the *San Felipe*, a Spanish
galleon filled with spices from the East, he found the ship equally laden with
pests, "a wonderful company of winged Moths, but somewhat bigger than
ours and of a more swarthy color," according to Mouffet. Drake probably
loaded some of these foreign roaches aboard his vessel the *Golden Hind*. In
1580, the year that the *Golden Hind* returned to its home port at Plymouth,
England, these "swarthy" insects from the genus *Periplaneta* became part of
the local invertebrate fauna.

Representatives of other cockroach genera were given free passage to far-
off lands by slave ships departing from the West African coast. When these
ships unloaded their human cargo in Trinidad and other New World outposts,

their holds were often refilled with tropical fruit, spices, and other treasured commodities. Concealed in these goods were the egg capsules, young, and adults of several roach species, which would soon become acclimated to European homes and markets.

The circuitous route taken by one of these shipboard species, the strikingly patterned harlequin cockroach (*Neostylopyga rhombifolia*), was well documented in 1945 by James A. G. Rehn, curator of insects at the Philadelphia Academy of Natural Sciences. In *Man's Uninvited Fellow Traveler—The Cockroach* (still the best source of information on blattid dispersal patterns), Rehn showed how this species was introduced to the port of Acapulco, on the west coast of Mexico, by Spanish galleons from the Philippines. Cargo from these ships was brought ashore, then carried overland to the Atlantic, where it was reloaded on other galleons bound for Spain. In this way, Rehn asserted, roaches were transported three-quarters of the way around the world, then given every opportunity to spread out even more. The range of the harlequin cockroach now extends from Malaysia to the east coast of Africa, with odd, localized hot spots in such places as Nogales, near the Arizona border in the Mexican state of Sonora.

Cockroaches continued to plague ships' passengers, cargoes, and crews throughout the eighteenth and nineteenth centuries. These insects were "in immense profusion, and had communication with every part of the ship, between the timbers or skin," wrote R. J. Lewis, an entomologist sailing from England to Australia in 1830s. He added:

> The ravages they committed on everything edible were very extensive; not a biscuit but was more or less polluted by them, and amongst the cargo, 300 cases of cheeses, which had holes in them to prevent their sweating, were considerably damaged, some of them being half devoured and not one without some marks of their residence.

Cockroaches were no less rapacious to the crews aboard clippers and steamships. They "drive sleep away by their pestilent odour, and their continual crawling over the face and limbs of those who are vainly endeavoring to seek repose," declared Reverend J. G. Wood in his *Illustrated Natural*

A bit of folk wisdom: "Poke root, boiled in water and mixed with a quantity of molasses will kill roaches in great numbers."

History, published in London in 1863. Even worse, professed Vernon Kellogg, author of *American Insects*, in 1908:

> Ships come into San Francisco from their long half-year voyages around the Horn with the sailors wearing gloves on their hands when asleep in their bunks in a desperate effort to save their fingernails from being gnawed off by the hordes of roaches which infest the whole ship."

Today nobody's playing *Victory at Sea* where the Blattaria are concerned. According to Drs. Louis M. Roth and Edwin R. Willis, coauthors of *The Biotic Associations of Cockroaches*, more than twenty thousand have been taken from one ship's stateroom, and about two thousand American cockroaches per hold and as many as twenty-four German cockroaches per cabin have been identified by pest control operators aboard the S.S. *William Keith*, a steamer docked in San Francisco Bay, after a ten-month voyage to the South Pacific.

Coffee, tea, or cockroaches?

With the same ease with which they took to sea, cockroaches have extended their home ranges from the air. The usual suspects—American, German, and Oriental cockroaches—are among the two dozen or more species that have overrun airplane cockpits, cargo holds, and galleys. Researchers Louis M. Roth and Edwin R. Willis have listed twenty recognizable species plus a handful of unidentified specimens removed from aircraft in Brazil, Puerto Rico, New Zealand, the Hawaiian Islands, and Sudan. These insects seem to know that even on the most modest single-prop Cessna, there are sufficient quantities of old chewing gum and crumbs from the pilot's sack lunch to make flying the friendly skies worthwhile.

Where airplane food has been scarce in the past, roaches have sought sustenance from the glues and dope used in wing construction. Fortunately for all concerned, these materials have been discarded by today's airplane fabricators, who work primarily with fiberglass and lightweight metal. But cockroaches continue to chew through electrical wiring and hull insulation, making them undesirable passengers on any aircraft.

"Mrs. Smith takes it amiss when you ask permission to collect 'roaches' in her house, and will prove to you any day the conspicuous absence of these unwelcome guests in the scrubbed and spotless pantry and kitchen. But with a candle go stocking-footed at night into the same kitchen and you will not unlikely find 'good hunting.'"

—Vernon Kellogg, *American Insects* (1908)

Cockroaches in space

Even outer space has done little to impede the progress of hearty roach pioneers. It's been suggested that at least one of these uninvited explorers has already set foot on other worlds.

During a preflight check of the Apollo XII command module, *Yankee Clipper*, a worker at Cape Kennedy spied a single cockroach inside the capsule and noted the occurrence in the mission log. The worker's report was designated an "open item" on the Flight Readiness Review, where it awaited further investigation. Only after the *Yankee Clipper* was well on its way to the Moon did anyone remember the open item.

"I drew a cockroach on a note card, and held it up to the television camera lens inside the command capsule," Commander Charles "Pete" Conrad Jr. told readers of the July 1984 issue of *Houston City Magazine*. "I think I told them [Mission Control in Houston] that we had resolved that open item in the log."

The final fate of this intrepid insect remains a mystery. It may have abandoned the *Yankee Clipper* prior to liftoff. Hermetically sealed in with the capsule's occupants, the roach might have traveled to the Moon and back, securely wedged in some small crevice in the craft's instrument console.

"Or perhaps," suggested *Houston Magazine*, "when the Apollo XII astronauts docked with the lunar module *Intrepid* prior to descending to the surface, the roach sneaked aboard the LM [Lunar Module] through the open hatch and exists, in some form, on the surface of our nearest neighbor in space."

An endangered cockroach

First described in 1951 by J. W. H. Rehn (son of James A. G. Rehn) as a "dark fuscous to blackish" Blaberid, the Tuna Cave cockroach, *Aspiduchus cavernicola* is a good-sized (one-and-three-quarter-inch to two-inch) resident of Cueva Convento, a network of caves in southern Puerto Rico's Guayanilla area. The Tuna Cave cockroach's limited distribution and its exposure to possible threats from residential and industrial development have inspired biologists with the U.S. Fish and Wildlife Service to include this animal as a Category Two candidate species under the federal Endangered Species Act.

According to Susan Silanger of the U.S. Fish and Wildlife Service's Caribbean Field Office in Boqueron, Puerto Rico, Category Two candidates are those for which "listing as threatened or endangered may possibly be appropriate, but for which conclusive data on biological vulnerability and threat is currently not available to support the listing." With recent revisions to the Endangered Species Act, she noted, the Tuna Cave cockroach was reclassified as "species at risk." This unique insect is currently protected by the Puerto Rican Wildlife Law (Law No. 70). Like many other tropical invertebrate species, the Tuna Cave cockroach could vanish from our planet before its ways are studied or its contributions to the regional ecology thoroughly understood.

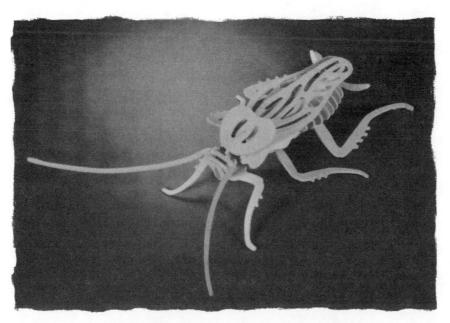

The fully assembled roach puzzle by Safari is anatomically accurate.

Blabeus giganteus, the Great Flier

How They Affect Our Lives

WHAT GOOD ARE COCKROACHES? This question implies that these creatures were created to serve humankind in some way. As roaches predate us by 340 million years, it's preposterous to presume that such servitude could have been written into the grand scheme of things.

"When we try to pick out anything by itself, we find it hitched to everything in the universe," observed the great American conservationist John Muir. With this in mind, it may be more valuable to ask in what ways these animals are "hitched" to us all.

In tropical settings, cockroaches play several key roles. They act as six-legged sanitary engineers, removing and recycling dead plant and animal matter from the forest floor, sharing this responsibility with worms, land snails, and representatives from several other invertebrate groups. It's been estimated that in the forested swamplands of Amazonia, for instance, members of the genus *Epilampra* turn over as much as 5.6 percent of the yearly leaf fall, returning important nutrients to this region's typically thin soils.

Cockroaches are also important sources of food for reptiles, amphibians, fish, birds, other insects, and a number of small and not-so-small mammals such as monkeys and bats. The bodies of roaches and their insect relatives are nearly three times as protein-rich as those of chickens or other, more familiar, meat sources. Estimates of the protein content of crickets has ranged from 24 percent (wet weight) to about 60 percent (dry weight). It's safe to assume that cockroaches are no less protein-rich.

Tropical blattarians may perform a third task, according to Donald Perry, an investigator of life in the forest overstory. His studies of one species—the bright yellow, black, and brown canopy cockroach *Paratropes bilunata*—indicate that this animal actively pollinates several overstory plants, including *Oreopanax*, an arboreal bush. A second species, the pale-headed cockroach (*Latiblatta lucifrons*) of the Huachuca and Santa Rita mountains of southern Arizona, may be an important pollinator of desert-blooming yucca plants (*Yucca sp.*), in whose pollen it wallows while feeding.

Recognizing the importance of roaches in maintaining natural systems, workers in Tucson, Arizona, intentionally introduced several species into the sealed environment of the Biosphere 2 project during the late 1980s. According to consulting entomologist Dr. Scott Miller of the Bishop Museum in Honolulu, Hawaii, both desert cockroaches and the Madagascan hissing cockroaches (*Gromphadorhina portentosa*) were included on the faunal lists of this futuristic Noah's Ark. Unfortunately, neither species was able to adapt to the artificial climate of Biosphere 2. Today, support staff wage war with some additional insect inhabitants, including the Surinam cockroach (*Pycnoscelus surinamensis*) and Australian cockroach—a pair of common greenhouse pests that somehow slipped into the fabricated forest during its final phase of construction.

Cockroaches in the laboratory

Despite their poor public images, cockroaches are the most frequently used specimens for the study of insect behavior, anatomy, and physiology. They also serve as reference subjects in broader areas of exploration, including pharmacology, immunology, and molecular biology.

From her study of laboratory-reared Madeira cockroaches (*Rhyparobia maderae*), Dr. Berta Scharrer determined that nerve cells secreted hormones into the blood. This and other findings spawned a new discipline, neuroendocrinology, which explores how the nervous system communicates with the body's endocrine system, affecting our early development and growth. It also earned Scharrer a nomination for the Nobel Prize in 1938.

Living and dead roaches are also widely used in elementary, high school, and college classrooms. To meet the demand for these popular insects, one catalog company, Carolina Biological Supply Co. in Burlington, North Carolina, routinely cultures as many as ten thousand American, five thousand German, three thousand Madagascan hissing, and four thousand giant and death's head cockroaches.

Cockroaches as animal chow

The enterprising Dr. Ivan Huber, professor of biology at Fairleigh Dickinson University, has also encouraged his students to investigate the feasibility of cockroaches as food sources for domestic animals. One of his students, Frances Marks, determined that German cockroaches could be anesthetized and ground up with a small amount of water to produce a souplike mixture. This glop was then freeze-dried and fed to laboratory mice. After eight days on this diet, the mice showed no ill effects from their food and had even demonstrated a gain in weight.

"Dietary protein levels for animals are based upon growth rate and feed costs, since an increase in protein content usually means an increase in cost," observed Marks. "Cockroaches, which can live on a wide variety of materials and have a high efficiency of food conversion would therefore seem to be an excellent source of food for an animal diet."

Culinary cockroaches

So what does a cockroach taste like? "Shrimp," according to British nautical scholar S. Melville-Davidson, who included this somewhat unnerving tidbit in a trade publication, *Some New and Interesting Points in Ships' Hygiene*, published in 1911.

"Salted cockroaches are said to have an agreeable flavour, which is apparent in certain popular sauces," offered I. A. C. Miall and Alfred Denny in their 1886 overview *The Structure and Life History of the Cockroach*,

Their extreme sensitivity to vibration could make cockroaches useful as earthquake predictors. In 1977, Dr. Ruth Simon of the U.S. Geological Survey monitored the activities of American cockroaches kept in boxes near three seismically active sites in California. Sensors in the boxes recorded a marked increase in cockroach activity immediately prior to earthquakes of small magnitude. The results of this one-year study were described as "not conclusive" but "very encouraging."

although just which sauces is not entirely clear. A second turn-of-the-century author went so far as to publish his recipe for making a sort of roach jam—a supposedly succulent dish made from a choice selection of blattarians that were to be simmered in vinegar all morning, then dried in the sun. The insects, freed of heads and intestines, were then supposed to be boiled together with butter, farina, pepper, and salt, producing a paste that could be spread on buttered bread.

A number of indigenous peoples, including the aborigines of Australia and the Lao hill tribe of Hua Hin, Thailand, have been known to collect and eat roaches raw. "The Laos in that district and in Korat will eat roaches, but in most other districts they are left alone and are said to 'stink'," added William S. Bristowe in his 1932 report to the Royal Entomological Society of London. "However, in all districts children appear to collect their eggs for frying."

A French entomologist in the 1940s, Dr. E. Brygoo, claimed to have known a commandant of the colonial army who ate cockroaches with evident pleasure, joining the Kissi (a French Guianan forest tribe) in this practice. Brygoo also wrote how the "more civilized" Annamites (the citizens of the Southeast Asian nation of Annam) also ate cockroaches, but only after they had been charbroiled over a flame.

Many folks may have unintentionally dined on cockroach chitin and flesh. "They are in everything, even the food," wrote A. N. Caudell in the journal *Entomological News*, describing his 1903 trip to the Canadian province of British Columbia. "On this trip I had them served to me in three different styles, alive in strawberries, a la carte with fried fish, and baked in biscuit."

In his book *Life on a Little-Known Planet*, writer Howard E. Evans described his discovery of what appeared to be an American cockroach, served as an unintended garnish on an order of beefsteak and onions at a small Texas cafe. Not one to be easily put off his feed, Evans calmly finished his meal, consuming everything but the large brown insect, which he left in a lifelike pose in the center of the empty plate. "The expression on the waiter's face when he picked up my plate was ample compensation for the health risk I took," he confessed.

> "Are bugs kosher?"
> "Uh, well, yes and no. There are references in the Bible that say six species of locust are kosher, but there's some discussion that people were really referring to locust beans."
>
> —Louis Sorkin, entomologist at the American Museum of Natural History, as interviewed by Patricia Volk ("An Entomological Study of Apartment 4A"), *The New York Times Magazine*, March 5, 1995.

Cockroaches in medicine

In ancient Greece, physicians routinely mixed cockroach entrails with oil of roses, stuffing this unsavory goop into their patients' infected ear canals. It should be noted, however, that Greek healers and natural historians were founts of misinformation. Pliny, for instance, proclaimed that a person carrying a woodpecker's beak would be immune to bee stings. For spider bites, his patients were advised to quaff a brew containing five live ants.

In the apothecaries of ancient China, dried roaches were prescribed to treat visceral diseases and maintain the digestion. Today they are still sold (for around two dollars an ounce) in the heart of San Francisco's Chinatown neighborhood. *Chinese Materia Medica*, a 1984 publication of the Orient Cultural Service in Taipei, lists several different kinds of blattarians that are believed to be beneficial in a wide range of situations—for feverish chills, swollen tongues, children "crying in the night with stomach ache," or for "breaking up retained blood clots, thereby promoting fertility."

Dried and powdered cockroaches were the main ingredient of *pulvis tarakanae*, a remedy for pleurisy and pericarditis that was embraced by doctors throughout Europe and the United States. This treatment may have originated in Czarist Russia, where a similar compound was manufactured for the treatment of dropsy. A few more medicinal uses of roaches (in this particular instance *Blatta orientalis*, the Oriental cockroach) were identified in the 1907 edition of *Merck's Index*:

> Constituents; Blattaric acid; antihydropin; fetid, fatty oil. Uses: internal, in dropsy, Bight's disease, whooping cough, etc. External; as an oil decoction for warts, ulcers, boils, etc. Doses: 10-15 grains in dropsy, as powder or pills, or four fluid drams decoction.

In an 1886 issue of *New York Tribune*, the curious medical practices in Louisiana were described. These included the prescribing of cockroach tea for tetanus, supplemented by a poultice of boiled roaches over the wound. Because of the "amazing" size of the Oriental cockroach in Louisiana, few would be required for a large plaster. The Blattaria were also fried in oil with garlic—a time-honored treatment for indigestion. Years later, the legendary

New Orleans jazz singer Louis Armstrong recalled being served a broth made from several boiled roaches, whenever he was ill. Whether this treatment soothed or caused Armstrong's gravelly voice has yet to be resolved.

For "dilation of the stomach," the 1930 translation of *The Medical Book of Malayan Medicine* recommends that roaches, "either 7 or 3 in number," be burnt and their ashes mixed with water. "Let the sick man drink this for three days in succession," it advises. "Do not inform him what his medicine contains, and let him be relieved."

Some stink

David Letterroach, from the Cockroach Hall of Fame

It wasn't by accident that *Blatta orientalis*, a common nuisance species of cockroach, was known as the "stinking Moth" in seventeenth century England. Many cockroaches, especially when they gather in large numbers, do in fact impart a peculiar, off-putting smell. "This is familiarly known as the 'roachy' odor," explained Glenn W. Herrick, no fan of domestic cockroaches, in *Insects Injurious to the Household and Annoying to Man*. "Dishes left standing on a shelf for some time where roaches are abundant are apt to become so impregnated with this odor that food afterwards cooked or served in them tastes unpleasant."

This heady mixture of scents emanates from the roach's feces, saliva, and waxy skin—all of which can be laden with highly aromatic sex pheromones. Referred to as "tartar of roaches" or "eau d' cafard," this odor is sometimes described as smelling of bitter almond—so much so, according to Julie Closson Kelley's *Little Lives*, that "one brought up in lands where the oriental cockroach is a kitchen familiar can never after enjoy that flavoring called 'pistache,' so dear to bakers of store cakes and makers of green frosting."

The anonymous author of *Natural History of Insects*, published in 1830, found the "roachy" odor far less likable, especially from cockroaches that "are very fond of ink and of oil, into which they are apt to fall and perish." Killed this way, these insects "turn most offensively putrid, so that a man might as well sit over the cadaverous body of a large animal as write with the ink in which they have died."

Cockroaches and disease

Considerable ink has been devoted to the roach's role as a carrier of diseases. The two most significant reports on this subject are *The Medical and Veterinary Importance of Cockroaches* by Roth and Willis (published by the Smithsonian Institution in 1957), and *Cockroaches—Biology and Control*, by Donald Cochran, a publication of the World Health Organization (WHO), revised in 1975. The two treatises list about forty different pathogens that are naturally carried by cockroaches and could potentially be transferred to humans and other forms of vertebrate life. These include such biggies as leprosy, bubonic plague, pneumonia, food poisoning, salmonella, and typhoid fever.

Laboratory studies have also shown that cockroaches may acquire, maintain, and excrete various viruses, making them likely suspects in the transmission of such life-threatening diseases as polio or infectious hepatitis. Proviral DNA, similar to components of human immunodeficiency virus (HIV), the causative agent of acquired immune deficiency syndrome (AIDS), has also been isolated from the genetic material of American cockroaches in central Africa, where this disease may have originated.

Cochran's report cites sixteen roach species that are considered "mechanical vectors" of organisms pathogenic to humans. As one might guess, all of these are domestic or peridomestic roaches—among them the ubiquitous German, American, Oriental, and brownbanded species. Tacked on to this list is a seventeenth species, the smokeybrown cockroach, whose presence "in both privies and homes would warrant suspecting it as an additional vector species."

The ways in which the blattarians transmit diseases are fairly straightforward: pathogens become attached to a cockroach's cuticle and they are then sprinkled on any object the animal walks across or otherwise contacts directly. Any roach that spends some of its time in a sewer pipe is, in all likelihood, an unknowing participant in this form of transmission. Internal organisms are just as easily passed along to any creatures that eat these disease carriers, either intentionally or accidentally. Numerous studies have produced fairly convincing information on the role of roaches in sharing certain nematode worms and other internal parasites with those that dine on their flesh.

"In 1948, the [polio] epidemic in Los Angeles County had many features pointing to an insect transmission rather than a direct transmission by bite of the biting stable fly, but had shown on several occasions: (1) that house flies contained the virus and dropped it in their feces; (2) that the virus could be picked up from human excreta; (3) that a monkey fed on bananas polluted by flies fed on virus, had given evidence of infection; (4) that the blow fly *Phormia regina*, carried the virus from two to three weeks; and (5) that the virus could be retained in cockroaches for 12 days"

—from *The Deadly Triangle: A Brief History of Medical and Sanitary Entomology* by William Dwight Pierce (1974)

Can cockroaches cross-contaminate themselves? Of course they can—at least in one University of North Carolina laboratory. In an elaborate set of experiments at this facility, a single salmonella-carrying roach was placed in an aquarium with food, water, and ten other roaches. After twenty-four hours, the contaminated insect was removed, and the other ten were tested to see if they had contracted the disease. The marked cockroach was then placed in a fresh aquarium with ten more of its buddies for another twenty-four-hour trial period. The process was repeated two more times, then the results were tabulated. Researchers found in the first tank that all but one of the cockroaches tested positive for salmonella. However, in each subsequent tankful, the number of infected insects was lessened. Of the ten occupants of final aquarium, only five were infected, suggesting that individual disease carriers become less contagious as time passes.

"Perhaps the most disgusting and potentially dangerous features of roach behavior are their habits of regurgitating some of their partially digested food and dropping feces, often at the same time they are feeding," Cochran has suggested, emphasizing a less-obvious but possibly more-direct means with which cockroaches can spread disease. It would not be surprising if sales of shrink-wrap and air-tight plastic food containers skyrocketed with the release of his report.

While it's widely recognized that cockroaches can carry disease, and that these diseases can be transmitted in several ways, the burden of proof, according to Cochran, still rests on the prosecution. To date, the most incriminating evidence of their disease-carrying capacity is still largely circumstantial. In 1963, following a rigorous roach-control program instituted at 124 buildings of several southern California housing projects, researchers noted a marked decrease in the incidence of infectious hepatitis. Hepatitis rates remained unchanged in nearby projects that did not receive pest treatments.

Without a smoking gun, no reputable attorney would consider filing a civil suit against cockroaches. Still, if you are reading this book at the breakfast table, and one of these insects happens to be crawling across your Cream of Wheat...

> "On every dish the booming beetle falls
> The cockroach plays, or caterpillar crawls;
> A thousand shapes of varigated hues
> Parade the table or inspect the stews.
> When hideous insects every plate defile;
> The laugh now emply and how forced the smile."
>
> —Andrew Nelson Caudell
> Notes on some Orthoptera from British Columbia
> (1904)

Cockroaches and the law

This is not to say that cockroaches haven't had their share of days in court. In 1970, the Georgia appellate court reviewed the testimony of a guest at Stuckey's Carriage Inn. While attempting to get rid of a roach, which had crawled up her thigh, the guest had become entangled in a bedspread, causing her to stumble and fall over a chair. The court rejected the owner's original contention that, even if the bedspread was negligently placed, the guest had an equal knowledge of the defect, and therefore should not receive damages. Instead it found that "the experience of the guest when she felt the cockroach crawling up her thigh, with its claws clinging to her skin, was a moment of stress of sufficient magnitude" to warrant the recovery of damages.

Many other cases have established the liability of landlords to the tenants of their roach-infested residences. In Leo versus Santagada, a tenant argued that seeing a dozen cockroaches within forty-five minutes was sufficient reason to abandon a recently rented basement apartment. The court denied the claim, "because the tenant and his family had made no attempt to help themselves by using methods that an ordinary housewife would use under the same circumstances, and the tenants did not notify the landlord in order to afford him an opportunity to remedy the situation." Alternatively, if an attack by "the verminous enemy" is sufficiently serious and comes from a source within the jurisdiction of the landlord, it violates the implied premise that a house or apartment will be habitable.

In general, renters are advised to consider the case of Pomeroy versus Tyler (1887), cited by Berenbaum in *Bugs in the System: Insects and Their Impacts on Human Affairs*. For this precedent-setting case, the court ruled that roaches, as well as water bugs, red ants, and bedbugs, should be no surprise to anyone renting a flat in New York.

Allergies and asthma

For their roles in promoting allergies, and asthma, cockroaches are guilty as charged. According to the National Institute of Health, as many as ten million to fifteen million Americans may suffer from roach-related allergies. This

"My husband trained roaches to attack me," Maude Kelly told tabloid journalist Louis Martin in the September 24, 1995 issue of *The Sun*. However her estranged husband, Bill, an entomologist, refuted the claim. *The Sun* described how experts, hired by Maude's lawyers, had detected roach pheromones on her clothing. This supported their client's assertion that Bill was using insects to drive her out of their suburban Pittsburgh home. The outcome of Maude's divorce hearing—and whether Bill had indeed unleashed trained combat cockroaches in his wife's bedroom and elsewhere— was never reported by *The Sun*.

A handsome hand-carved cockroach kitchen magnet
(from the collection of Steve Kutcher).

makes cockroaches the second greatest offender, after dust. In addition, there is reason to believe that an initial allergic reaction to roaches may cause subsequent allergic reactions to other hard-bodied invertebrates. Diners with these compound allergies would be robbed of delicacies such as lobster, crab, or shrimp.

Allergic reactions occur when our immune system reacts inappropriately to perceived threats in the environment. In the case of cockroach allergies, these threats are tiny particles of protein—the excrement, shed skins, partially consumed food, and pheromones produced by domestic roach species. Some of these proteinaceous products are extremely long lived; they resist boiling water, harsh chemicals, and ultraviolet light, and they can remain allergenically potent for decades.

In response to this assault on the senses, the immune system triggers the release of histamines. These complex carbon compounds go to work on veins and arteries, which, in turn, has different effects on the blood supply to various parts of the body. Mild reactions to roach allergen could result in a runny nose. Or it might cause the skin to break out in a rash. Labored breathing and other symptoms of asthma are not uncommon, and in rare instances, severe reactions may lead to death from shock.

Sufferers of cockroach allergies are more commonly found in low-income housing and inner-city apartments with inadequate roach control programs. Both human and insect occupants of such residences are often crowded together—a code red condition for allergic individuals. Cockroach allergies may be a bigger problem in the northern states, where long, cold winters keep roaches and people cooped up in tight quarters with closed windows. However, people in the southwest and south-central states appear to be more prone to allergies—not just to the blattarians, but to nearly all of nature's aromatic substances—because of the warm, moist environment, which is favorable year-round for the plants and animals that trigger allergic reactions.

Creepy cockroaches

What is it about cockroaches that gives so many of us the heebie jeebies? One answer to this question lies in the dark recesses of the subconscious mind. Some scientists have theorized that this distaste is instinctive—a by-product of our instinctive mistrust of stinging insects, biting spiders, and venomous snakes. Thus people are primed to react negatively to roaches or any other critters that resemble these potentially dangerous animals.

We are more easily frightened by animals that are ugly, slimy, fast-moving, and/or unpredictable, a related theory suggests. To test this hypothesis, participants in one British psychological study were asked to rate these very attributes in five different invertebrates—a moth, grasshopper, spider, butterfly, and beetle. As one might anticipate, the spider was the unanimous winner. However, the beetle also received surprisingly high scores in ugliness, sliminess, and speed, placing it second on the list. It's safe to assume that, had a giant cockroach been added to the study's lineup, it would have given both the spider and the beetle a run for the money.

Equally insidious are the repeated oral warnings of parents or peers, which can only reinforce a latent aversion to creepy-crawlies. Data suggests that such reinforcement is especially prevalent among women: mothers are more prone to pass their insect fears to daughters than fathers to sons.

"Children up to the age of about four years have no aversion to cockroaches, but are quickly taught by parents that cockroaches are filthy and

should not be touched or put into their mouths," wrote Philip Koehler and Richard Patterson. In their 1992 paper, the two U.S. Department of Agriculture researchers cited a behavioral study in which rubber cockroaches were placed in kids' drinking glasses. Children less than four years old would readily sip from these vessels, whereas children over this age refused to put them to their mouths. "Such studies readily demonstrate that cockroach aversion is learned, not inherited," concluded Koehler and Patterson.

Delusions of grossness

A persistent, irrational, or unreasonable fear of insects is called entomophobia—a mental state akin to the more widely publicized arachnophobia, or fear of spiders. In severe cases, this fear manifests itself as delusory parasitosis, an unshakable belief that live organisms, primarily insects, have parasitized the skin. In a second, closely related illness (known as delusory celptoparasitosis), the imagined insects have infested homes or workplaces, making them uninhabitable.

In their 1963 paper, Albert H. Schrut and William G. Waldron identified several commonly reported symptoms of the two illnesses:

> Imagined "bugs" are usually black or white when first noted, but, later on, may change color.

> "Bugs" may emerge from common household items, such as toothpaste, petroleum jelly, or cosmetics.

> In homes, a supposed infestation may become so severe as to literally force the afflicted person to move to another location.

> The supposed infestation may have lasted for two or three months—far longer than most genuine insect, louse, or spider infestations typically last.

> A patient's description of the infestation is often so compelling that other, non-afflicted family members will support the decision to move.

> "I have seen large bags (100 pounds) of both onions and potatoes emptied on the produce table at a time when the number of roaches running out of the bag surpassed the number of onions or potatoes in the bag."
>
> —D. M. Delong,
> *The Supermarket's Problem*
> (1948)

In nearly all instances investigated by Schrut and Waldron, a real invertebrate infestation appeared to precede (and possibly trigger) the delusory insect attacks. The two authors described one case in which a family repeatedly contacted the health department claiming that roaches, bedbugs, and lice infested their house. However, a low-grade population of American cockroaches were the only pests observed in the home. Husband and wife also claimed that a "poisonous substance" from a newly painted room was causing illness to appear in members of their family.

"All four members of the family said they saw the insects frequently and claimed symptoms of weakness and headaches from the 'poison paint' and insect bites," wrote Schrut and Waldron. "Investigation revealed a great emotional struggle between the mother and father, which apparently precipitated a psychotic episode in the mother." Faced with threats of separation or divorce, the mother fixed the blame on the parasites and poisonous paint. This involved abandoning the house temporarily, taking the children with her, and leaving her husband, "thus separating herself from the real culprit involved in her unconscious."

> *"An Israeli housewife's battle with a stubborn cockroach landed her husband in hospital with severe burns, a broken pelvis and broken ribs"*
>
> —JERUSALEM POST

Don't try this at home

On August 25, 1988, the Reuters News Bureau in Tel Aviv picked up a shocking roach tale and transmitted it to newspapers around the world. "An Israeli housewife's battle with a stubborn cockroach landed her husband in hospital with severe burns, a broken pelvis and broken ribs," this memorable story began, citing the *Jerusalem Post* as its source.

The battle began when the wife caught a roach in her living room, stamped on it, and then tossed it in the toilet, subduing the insect with an entire aerosol can of insecticide. Additional fighting broke out when the woman's husband returned from work and threw a still-burning cigarette butt into the toilet bowl, igniting the insecticide vapors and "seriously burning his sensitive parts."

"Ambulancemen, shaking with laughter at the incident, dropped the stretcher down the stairs, causing the unidentified man further injuries before finally getting him into hospital," the story concluded.

Six days after the story broke, the *Jerusalem Post* printed a retraction, explaining that "a good tale got so tangled in the telling that it assumed a newsworthiness it should never have had." The *Post's* retraction has done little to kill the story's status as a classic urban legend.

Frightening but fun

Although cockroaches are upsetting to some people, they are entertaining to many others. Capitalizing on our feelings of revulsion and fascination for these animals, several filmmakers have cast the Blattaria as leads in low-budget horror. George Romero's *Creepshow* ranks as a classic, largely because of its concluding vignette (see page 28). Also worth viewing, if just for the laughs, is the low-budget shocker, *The Bug*, directed by Jeannot Szwarc in 1975. Even more astoundingly cheesy is *The Nest*, by Terence H. Winkless, which pits genetically engineered killer cockroaches against some juicy and flavorful humans ("Roaches have never tasted flesh... until now" announces the poster for this 1987 film).

Roaches have made cameo appearances in dozens of mainstream motion pictures. They earned Julie Andrews a free meal in *Victor Victoria*, and got Steve McQueen through some hard times in *Papillon*. They disrupted a fight between Dianne Keaton and Tuesday Weld in *Looking For Mr. Goodbar*, and harassed Harrison Ford in *Indiana Jones and the Temple of Doom*.

As of this writing, the filming of *Joe's Apartment*, a feature-length picture based on a popular three-minute MTV video segment about a roach-ridden flat has begun. With filmed sequences that include an Esther Williams-like blattarian water ballet in a toilet bowl and a bronco-busting roach going for a ride on the back of a cat, this promises to be the most elaborate vehicle for cockroaches in the history of the silver screen. During the making of this epic the film's producer, Diana Phillips, become a stand-in for one of the stars, allowing hundreds of roaches to scramble across her chest.

Cockroach-abilia

Archie McPhee, the self-proclaimed "Outfitter of Popular Culture," has all but cornered the market in cockroach accessories. Since 1980, the Seattle, Washington, mass marketer of kitsch has sold over two million rubber roaches. This figure includes the four-and-a-quarter-inch Deluxe Jumbo Roach. But it does not include McPhee's most recent catalog addition: a water-soluble roach "egg" containing ultra-absorbent plastic nymphs.

The giant-sized replicas are most popular in Florida and Texas, where people brag that their live roaches are the same size. The regular-sized, one-and-three-quarter-inch brown rubber roaches and their glow-in-the-dark counterparts appeal more to New Yorkers, who give them as housewarming presents.

Also popular with New Yorkers has been the American cockroach hand puppet made by Folkmanis, Inc., of Emeryville, California. These artfully crafted soft sculptures have a tan, corduroy body, glossy Naugahyde wings, a pair of nine-inch-long leather antennae, and a six-fingered glove for legs. Two dabs of brown paint on its nylon pronotum are solid clues of this puppet's membership in the genus *Periplaneta*.

For sheer impact, there's nothing better than the human-sized cockroach costumes fabricated by Foam Domes in Minneapolis. These creations are intended for use by educators and interpretive naturalists, so the patterns (which sell from fifteen to twenty-five dollars) focus on the good, the bad, and the ugly of the animal kingdom. Ready-to-wear cockroach costumes are priced at two hundred fifty dollars. In addition to their cockroach costumes, current subjects include a monarch butterfly, a bat, and an anatomically correct ladybird beetle—suitable attire for winning over the most devout entomophobe.

An orthopteran anthem

If there were such a thing as a national anthem for roaches, it would have to be *La Cucaracha*. More than thirty versions of this lively tune have been recorded in this century by such musical luminaries as Charlie Parker, 101

Strings, Louis "Satchmo" Armstrong, Ferrante, and Teicher. Chicago blues master Big Walter Horton transformed it into a searing harmonica solo, and Bill Haley and the Comets managed to give it a rockabilly beat.

The song was written around 1910, at the start of the Mexican Revolution, by the high-spirited followers of Francisco "Pancho" Villa during his campaign. Its title refers to much more than blattarians, as the word "cucaracha" has a second, slang meaning—a "dried-up old maid"—which was also the nickname of Villa's chief rival, General Venustiano Carranza.

There are countless verses invented on the run, as it were, by the soldiers of Villa's army and their followers. The best known of these is also its most enigmatic, translated from its original Spanish as follows:

The cockroach, the cockroach
Doesn't want to walk anymore
Because she hasn't, because she lacks
Marijuana to smoke.

Some of these verses are really commentaries on the Revolution, which lasted roughly ten years:

The Carranzistas are already going,
They are going to Laredo.
They are no longer Convencionistas
Because they are very frightened.

Other verses contain droll observations about life:

When a fellow loves a maiden,
And the maiden doesn't love him,
It's the same as when a bald man
Finds a comb upon the highway.

Archie McPhee's rubber roach—
a brisk seller.

A few artfully combine the two perspectives:

> *One thing makes me laugh—*
> *Pancho Villa without a shirt.*
> *Now the Carranzistas are leaving*
> *Because the Villistas are coming.*

Louis Armstrong made up his own verse, scat-singing it in 1935:

> *La cucaracha, la cucaracha*
> *Sing no matter where you are;*
> *La cucaracha, la cucaracha*
> *Try it on your old guitar.*

Children today learn a different version at school:

> *When they dance the cucaracha,*
> *And I hear the music playing,*
> *To the plaza then I hurry,*
> *Join the dance without delaying.*

The original lyric, referring to the roach's penchant for pot, has been deleted. Instead, students learn that the cockroach can't walk because it lacks *las dos patitas de atras*—its two little hind legs!

Above: Design for a cockroach costume by Laura Emmers of Foam Domes.

Sex, Food, and Death

This Is How
It All Begins . . .

THE START OF A COCKROACH is when a sperm cell and an egg unite. However, a lot goes on before these two packages of chromosomes meet.

Cockroach courtship is typically initiated by the sexually mature female. She does this by flexing her lower body, raising both pairs of wings, and assuming a pose known as her "calling stance." From this position, she secretes a powerful chemical attractant, the insect equivalent of French perfume, from special membranes on her back. This attractive (in the most literal sense of the word) position is illustrated at left (the drawing shows a female *Xestoblatta* assuming the classic calling stance.

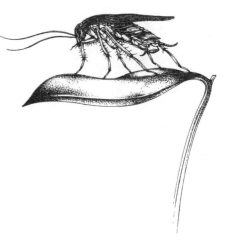

This "love potion" is made of hormonelike substances, called pheromones—interwoven chains of carbon, hydrogen, and nitrogen. Measured at less than a billionth of a gram for at least one species, the brownbanded cockroach, the pheromonal product of one female is quite volatile. Carried on the wind, it sends a subtle sexual invitation to any male roaches in a radius of at least thirty-three feet.

Sensors in the males' scent-receptive antennae get the chemical message, encouraging their bearers to seek out the female transmitter. In tropical habitats at dusk, male roaches take higher perches on bushes and shrubs—a move that puts them in better positions to sample the mists, sniffing for prospective mates.

Fencing for foreplay

When sexually receptive adult male and female cockroaches meet, they engage in a brief bout of antennae fencing. During this ritualistic act, the insects face off, then lightly but rapidly lash each other with their whiplike antennae. The many sensory receptors on these "sniff-whips" start to tingle, giving the two roaches a rush of tactile and chemical stimuli.

After a minute or two of this electrically charged foreplay, the male roach takes the lead. First, he turns his back on the female. Then he curves the tip of his abdomen downwards, bends his legs to lower his head and thorax, and raises his wings to an angle of about sixty degrees. This is his way of showing off his *excitator*, an appropriately named spot on his back. It is from this spot—a small, bristly lobe on the seventh abdominal tegite—that a second sex pheromone is released.

The male's sexual scent has an equally apt name, *seducin*. It has an immediate effect on the female, who steps forward and starts to nuzzle the male's excitator.

Cockroach Kama sutra

With the female now on his back (and, in some species, with her legs wrapped around his middle), the male pushes backward. Arching his abdomen upwards, he extends his genitalia in an attempt to grasp the genitals of his partner.

At the actual moment of genital contact, the female is standing atop the male, facing his head. But as soon as the hookup is completed, she steps sideways off his back, turning 180 degrees away from her mate. Now facing in opposite directions, the pair will remain linked for an hour or more—as long as it takes for the male's packet of sperm to be passed. This oval-shaped white packet is called a spermatophore. Inside it are all the sperm cells a female will need to produce her first batch of fertilized eggs. In several species, successive batches can be inseminated by sperm from the same spermatophore. In others, a fresh packet must be provided.

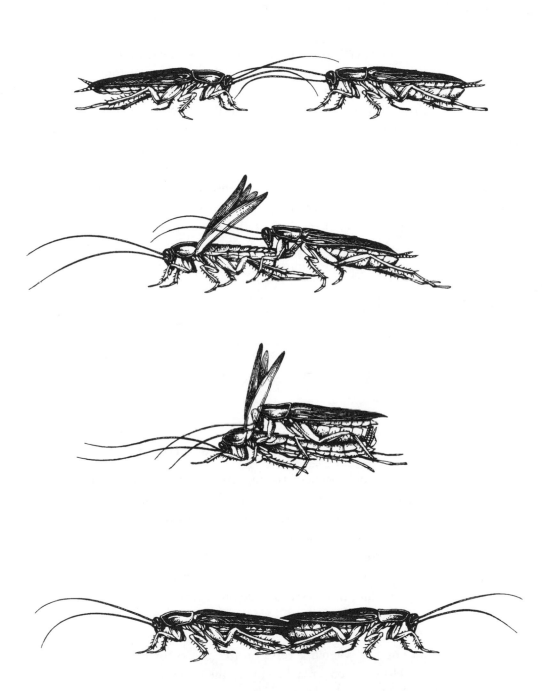

Cockroach sex (top to bottom) includes fencing, wing-raising, nuzzling, and hooking up.

The spermatophore is filled with supplemental sources of nitrogen called urates. Several hours after copulating, when all the sperm has been drained from this packet, females of several species push the spermatophore out of their bodies. They then devour its urate-laden remnants. Refortified, these females can now comfortably carry their fertile eggs to term.

Of course, not every one of the four thousand species of roaches follows this exact sequence of events while courting and copulating. Male American cockroaches, for example, dispense with the formalities and charge at sexually mature females. They thrust their genitals at the females', trying to grab hold. Male cockroaches of the genus *Blaberus* excite their loved ones by butting them with their heads. Others perform for their partners, singing and dancing with great intensity.

Sonographs of cockroach songs—pulses and bursts (from a paper by Hartman and Roth)

> The expression "sound as a roach" has absolutely nothing to do with the cockroach. Linguists trace this odd phrase to St. Roche, patron saint of prisoners and hospital patients. After a statue of this saint was said to have quelled an outbreak of the plague in the year 1414, people began praying to St. Roche, asking him to make them sound, sound as the miracle worker himself.

Love songs

Yes, cockroaches can sing. Admittedly not like Mario Lanza or the Beatles, but more like gypsy violins. With insects, this singing is called stridulation: the act of rubbing together two body parts, usually the legs or wings, for the sole purpose of sonification.

Crickets and katydids are probably the best known stridulators, filling the night air with their pastorale songs. "I love to hear thy earnest voice wherever thou art hid, thou testy little dogmatist, thou pretty Katy-did," offered the poet, Oliver Wendell Holmes. A few classical composers, among them Josquin des Prez, have incorporated the sounds of crickets into orchestral works. To date, no one has written a verse or a concerto for the stridulating roach.

How do cockroaches make their music? Run your fingernail across the teeth of a hard rubber comb and you'll get the basic idea. Small filelike ridges (called striae) along the edges of a roach's pronotum take the part of the comb's teeth. The thick outer edge of the forewing acts like the fingernail. The rest of the forewing becomes the cockroach's PA system. Its paper-thin surfaces amplify the tune and transmit it to the rest of the world. To raise or lower the pitch, the cockroach simply changes the pace of its body parts rubbing together.

Like songbirds, humpback whales, and human beings, male Madeira cockroaches sing to attract members of the opposite sex. Each singer has about 50 striae on its pronotum and between 700 and 900 striae on its forewing. Females have slightly less. Both sexes also stridulate when alarmed, producing what entomologist Ashley B. Gurney has characterized as "a low, squeaking sound." Emanating from an insect no bigger than a clothes moth, this low squeak is just enough to startle a hungry lizard or bird, causing it to drop its pale brown catch. To learn how the Madeira cockroach produces its squeaky sound, Gurney actually took several dried specimens to his laboratory. Here, he manipulated their body parts with his fingers—in effect, serenading himself with one of the world's smallest violins.

Both male and female cinereous cockroaches (*Nauphoeta cinerea*) stridulate when disturbed, sending out a string of high-powered chirps in excess of sixty decibels—just a few decibels less than an alarm clock. Only the males stridulate for sex, and only after all other courtship strategies have failed. In the close proximity of a sexually mature but unreceptive female, a male cinereous cockroach will stiffen its legs, pump up its abdomen, and start to vibrate, rapidly moving its pronotum like a guitar pick, forward and back, side to side, against the striae of its front wings.

From all this activity comes the male's courtship call—which can be summed up as "two to six complex pulse trains," according to H. Bernard Hartman and Louis M. Roth, "followed by a long series of disyllabic chirps." These pulse trains and chirps are combined into phrases of five or ten seconds each. These can be strung together to form "sentences" lasting as long as three minutes. Should the hard-to-please female move away from her singing suitor, this rather lengthy love song comes to a halt. Move over Madonna!

The little cockroach
Running all up and down
my house.
The little cockroach
Quiet as a little mouse.
He gets in trouble
Snooping here and
everywhere.
The little cockroach
Always keeps my
cupboard bare.
One day when something
was baking,
Wondered he what's in
the making.
'Cause it looked
so appetizing
To see the batter
slowly rising.
To the edge he
started skipping.
Found himself he
was slipping.
Into the batter hot
and blazin'
Now he's just
another raisin.

Lyrics to "La Cucaracha,"
as sung by Howard
"Louie Bluie" Armstrong.

Bang, rattle, hiss

Cockroaches sonificate in several ways. During courtship the tropical *Eublaberus posticus* produces a tapping noise by banging its abdomen on the ground. The Madagascan hissing cockroach constricts its abdomen to expel air from its abdominal spiracles, producing a double-barreled blast of nearly ninety decibels, audible from a distance of twelve feet. This behavior has earned the two to three-and-a-half-inch insect its nickname, the "blower," given by European settlers of the cockroach's east African island haunt. The indigenous people of Madagascar's central highlands call this species *Kofoko-foka*—an appellation that approximates the hisser's ratchetlike call.

Male and female Madagascan hissing cockroaches are easy to tell apart. An adult male's pronotum sports a pair of short, blunt horns, used as battering rams during territorial shoving matches. Females lack these protuberances, and their bodies are much larger and heavier than those of their mates. In the wild, both sexes live in large groups, with sexually mature adults and juveniles in various developmental stages scavenging side by side on rotten logs. Both sexes sound off when disturbed and will hiss even more vigorously to escape being eaten by predators.

Male Madagascan hissers (above) also hiss in three separate social contexts—during aggressive encounters with other males, throughout courtship, and while copulating. In one experiment, the specialized spiracles of male cockroaches were sealed shut with cyano-acrylate glue. No longer able to hiss, these poor unfortunates could still participate in normal courtship, but they were largely unsuccessful at mating with females. Only after researchers played tape-recorded courtship hisses to the females could the silenced males find partners that were willing to mount!

Misdirected amour

Male cockroaches may occasionally make amorous advances toward other males. It's thought that such attention may be inspired by molecules of female sex pheromone that, in crowded conditions, fasten themselves to other courting males. While rather uncommon, such homosexual behavior has been observed in a wide range of species, including German, American, Australian, Madeira, and cinereous cockroaches. In laboratory cultures, male cinereous cockroaches will even sing to each other when no receptive females are around.

Pheromone-driven impulses are easily explained. However, behaviorists remain puzzled by males that act like females, copying their mounting and nuzzling activities during courtship. Formally known as pseudofemale behavior, such maneuvers often initiate copulatory behaviors among the males being mounted in this way, but they rarely result in genital hookups.

While seeking an explanation for pseudofemale behavior, Peter Wendelken and Robert Barth Jr. watched the precopulatory antics of four species in the genus *Blaberus*, before, during, and after copulation. They observed frequent outbreaks of fighting and chasing among males, usually initiated by unpaired males to disrupt the activities of their copulating counterparts. During these aggressive outbreaks, an attacking male would frequently clamber onto the back of its opponent, grasping and biting its tegmina or pronotum.

The pair of entomologists also found that, as a last-ditch effort to thwart copulation, the attacker would push a sex-driven female aside, taking her place on the back of an actively courting male. But unlike the female, this intruder would give the raised wings or the excitator a nasty nip, causing its victim to lurch forward in an attempted escape. Only by taking the female's place atop the male cockroach could a rival forestall copulation, giving himself one more chance to win over the female.

A do-it-yourself cockroach

When it comes to copulation, the Surinam or bicolored cockroach (*Pycnoscelus surinamensis*) is the exception that proves the rules. Females of this Indo-Malay species can reproduce without any input from males. However,

Male and female Pennsylvania wood cockroaches are so different in appearance that they once were considered separate species. Males have wings that extend past the abdomen; females are smaller and have much shorter wings.

there's a hitch to this method: only female cockroaches—as many as thirty-five at a go—are produced. Researchers have identified two separate strains of Surinam cockroaches, only one of which can reproduce in this fashion.

This method of reproduction, called parthenogenesis, is not all that unusual among insects. According to Frost's *Insect Lives*, one scientist reared ninety-eight generations of aphids parthenogenetically over a period of four years and three months, terminating this experiment only because of boredom. (It was possibly boring for the aphids, too.) Outside of the laboratory, aphids appear to capitalize on this asexual mode of reproduction only during summer months, when the weather is warm and food is plentiful. Embryos within these parthenogenetic insects' ovaries actually contain tiny embryos, and these, in an arrangement reminiscent of those wooden nesting dolls from Russia, have even tinier embryos inside of them.

Unfertilized eggs of German, brownbanded, and a few other roach species are known to undergo partial development. However, none of these eggs ever hatch. Unfertilized eggs of Oriental and American cockroaches have been reared to maturity in laboratories, but it's unlikely that this achievement could be replicated in the wild. For the most part, homeowners and renters with roach problems can find comfort in the fact that, as with birds, fleas, and bumblebees, it typically takes two to tango.

A nitrogenous gift

After copulation, males of some species award their mates with urates secreted from special uricose glands. On the tropical species *Xestoblatta hamata* (a distant relative of the German cockroach), these glands are located at the tip of the male's abdomen. Immediately after unhooking its genitals, a male will raise his wings and telescope his abdomen, directing his rear end at his mate. She then dines on the urates—a whitish goop from his genital chamber. The feasting can last for four or five minutes, according to researchers who clocked these animals in the forests of Costa Rica. With no females to eat it, the excess uric acid from urates can accumulate and eventually poison a male. For this reason, copulation is relatively protracted (around four hours in length), giving females ample time for their hunger for urates to build.

Why do female cockroaches crave urates in the first place? Laboratory tests support the theory that gravid females incorporate these nitrogen-rich products into their eggs before they are deposited. These serve as reserves for roach embryos as they develop inside their egg capsules. In this way, the male parent's investment of nitrogen benefits not only females but their progeny as well.

A convenient carrying case

Cockroach eggs develop inside a purse-shaped egg capsule, or ootheca. This capsule is made of a tough but pliable material not unlike cow horn, secreted from a pair of colleterial glands that open into the female's oviduct. A good description of how this works is contained in Miall and Denny's *The Structure and Life History of the Cockroach*:

> The secretion is at first fluid and white, but hardens and turns brown on exposure to the air. In this way a sort of mould of the vulva is formed, which is hollow and opens forwards towards the outlet for the common oviduct. Eggs are now passed one by one into the capsule; and as it becomes full, its length is gradually increased by fresh additions, while the first-formed portion begins to protrude from the body of the female.

The eggs fit into individual compartments within the ootheca. These compartments are arranged in two parallel rows. The number of eggs deposited in each ootheca (egg capsule) varies, according to the genus and species, as does the number of oothecae that one female can produce in a lifetime:

	NUMBER OF OOTHECAE	EGGS PER OOTHECA
American cockroach	10-15	16-28
Australian cockroach	2-3	22-24
Brownbanded cockroach	1-2	16
German cockroach	4-8	37-44
Surinam (parthenogenetic) cockroach	20	26

When the capsule is full, its open, upper edge is sealed shut. This produces a finely notched seam, called a keel, that runs the length of the ootheca. The keel contains tiny air holes through which the developing embryos can breathe. Both oxygen and carbon dioxide from respiration pass through a thin, transparent membrane that covers each egg. The organic equivalent of shrink-wrap, this membrane holds in moisture, protecting the developing embryo from drying out. This important adaptation may have assisted roaches in surviving the severe droughts of early geologic times.

What happens next to the egg capsule differs among the various cockroach families. American cockroaches and other members of the family Blattidae deposit their oothecae in safe places a few hours or days after they are formed. These parents play no roles whatsoever in helping their youngsters to hatch.

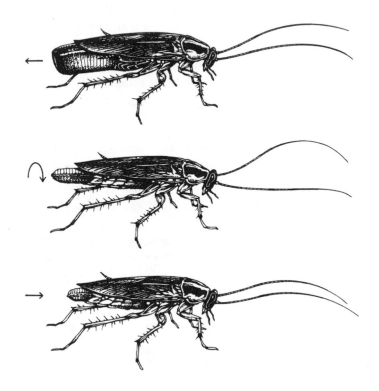

Staying slender while carrying eggs: extrusion, rotation, re-insertion.

Madeira cockroaches and many other members of the family Blattellidae carry their egg capsules to full term, conferring a greater degree of parental protection and care to their offspring. The young hatch inside their mother's womblike brood sack—a phenomenon technically known as "false ovoviviparity."

German cockroaches and several others in the Blattellidae carry their oothecae for several weeks, then jettison them a few hours before the eggs are due to hatch. These roaches have special tricks for keeping their compressed body shapes while they carry their purse-shaped capsules. They first extrude the oothecae three quarters of the way out of the body, then rotate it ninety degrees before drawing it back in.

Live-bearing cockroaches

A third strategy for bringing new cockroaches into the world belongs to the Pacific beetle or cypress cockroach (*Diploptera punctata*), a member of the Blaberidae. The female of this small, brown Indo-Pacific species carries its embryos in a womb of sorts, foregoing all but a remnant of the egg capsule. What's more, the lining of this womb secretes a substance that, for all intents and purposes, could be called "cockroach milk."

This glandular fluid is a nutritious beverage, composed of 45 percent protein, 25 percent carbohydrate, and 16 to 22 percent fat. It sustains the dozen or more embryos from the time they've developed functioning mouths and digestive tracts until they are born.

Straight from the womb, newborn Pacific beetle cockroaches are enormous (around five-sixteenths of an inch), nearly twice the size of the hatchlings of American cockroaches and a few other considerably larger species. Pacific beetle cockroach parents bestow on their young the same evolutionary advantage that we humans offer our children. Born bigger, they are better able to survive and pass their genetic materials on to their young. Such a rearing strategy is unique among cockroaches but it is common in other insect orders, most notably among the oddball aphids.

Birth of a blattarian

For the first two weeks there are no outward signs that anything is happening within the loaded and sealed ootheca. However, inside this capsule, the developing embryos are starting to look more like insects and less like slices of Wonderbread. In many species, a row of dark spots—one for each fetal cockroach—becomes dimly visible through the egg capsule's outer walls. At this stage in development, the fetuses have easily discerned eyes, mouthparts, antennae, and legs.

In as few as fifteen or as many as ninety days from the time they were first formed, the babies are ready to emerge. They start by pushing their way toward the egg capsule's narrow seam. Then they begin to gulp air, around two or three tiny bubbles per minute. The bubbles collect in their alimentary canals and cause each unborn roach's body to inflate, nearly doubling in size.

Crammed into neat and orderly rows, the rapidly expanding insects can only grow in one direction—out. The capsule's seam ruptures, and they spill out. It's still a bit of a struggle for them, for each emerging blattarian must pull itself out of the thin, membranous pellicle that enshrouds it, much like a mammal's amniotic sack. A few spasmodic twists and turns and these youngsters tear free—a process called an embryonic molt. They can now wave their antennae about and wiggle their six long legs, which until this moment have been kept tightly constrained.

A mere nymph

American cockroach:
female with nymph

Known as nymphs, newborn roaches are bloated from all the air they've just swallowed. But after a few minutes, they deflate, collapsing to around half their former length and girth. At first their bodies are almost transparent, and one can actually see the bright greenish hues of their innards. But an hour or two after emergence, the chitinous exoskeleton of the hatchlings begins to harden, gradually losing its ghostly pallor.

All newborn roaches are wingless and have noticeably shorter antennae and cerci than the adults. Their features enlarge and change as these animals pass through a succession of developmental stages, which are influenced by ambient temperatures and other environmental conditions. German cockroach nymphs undergo five to seven of these stages, each lasting from five to fourteen days. Thus it takes them fifty to sixty days to reach adulthood. Giant cockroach nymphs undergo nine to eleven molts, spread out over 257 to 277 days.

It takes nymphs of the dusky cockroach (*Ecotobius lapponicus*) and a few other temperate species even longer to reach adulthood. After lying dormant through the winter, these insects emerge from their egg capsules in early spring. They grow and change throughout the summer and most of the fall, then wait out a second winter before becoming adults. This final phase of a dusky cockroach's life is astonishingly brief—from July to September or possibly October of one year.

Mysteries of the molt

Each stage of a nymph's development is punctuated by a molt, during which these animals literally crawl out of their skins. This is the only way that roaches of any age can get any bigger, for their exoskeletons are made of rigid chitin and won't give an inch. So when it's time to start growing up, the nymph once again gulps down air. This causes its cuticle to split, from the head down to the tail, in a straight line along the back. From the shell of its former self, a new roach emerges.

This born-again cockroach is not a pretty sight. It is pale, skinny, and weak—in other words, a mess. Finding one of these half-baked critters in their pantries or shower stalls, people often mistake them for albinos or "freak" mutants. However, after capturing this "rare" specimen and keeping it in a Tupperware container, they soon realized they've made a mistake.

The freakish nymph starts to change almost immediately. It swallows even more air, causing its wrinkled, pigmentless skin to inflate like a balloon. The wrinkles disappear as the cuticle stretches in all directions at once. Still soft and supple, the freshly formed exoskeleton must cure for a while. It also acquires progressively darker colors—a process called "tanning." After a few hours this new outer covering is as tough as fiberglass. The fully reconstituted roach is now significantly larger than it was a few hours ago.

> If insects had the gift of speech, as we understand it, I am sure a main topic of conversation would begin: 'Let me tell you about *my* molt'"
>
> —Edwin Way Teale, *Near Horizons: The Story of an Insect Garden* (Dodd, Mead & Co., 1942)

Having completed its molt, the nymph gets down to brass tacks, eating its old exoskeleton, so that its protein and chitin can be incorporated into the new self. Sometimes people find these cast-off suits of armor before a cockroach has had the chance to finish it off. Remarkably thin, full-scale replicas of their original owners, these should be handled carefully and saved for later examination with a hand lens. Every detail of the living cockroach—from its antenna stalks to the tiny "portholes" that once marked the openings to its spiracles—is depicted in these frail souvenirs.

New legs for old

If you could grow a new skin every now and then, why would you grieve over the loss of a few limbs? Cockroaches often regenerate previously severed or broken legs, mouthparts, and antennae during the act of molting. Sometimes the regenerated legs lack a segment, making them shorter than the rest. However, in most instances, regeneration is an all-or-nothing affair.

The bodies of German cockroaches and many other species have at least fourteen well-defined points of weakness, where breakage most frequently occurs. When a limb is torn from one of these points, the severed connection quickly shuts itself off like a bulkhead to prevent any critical losses of body fluids.

Bringing up baby

Many roach species simply drop their egg cases and run, leaving their offspring to fend for themselves. However, in false ovoviviparous species, the young spend their first few hours near their mother. This way they can take advantage of the limited protection she affords while the chitinous cuticles harden. Fully armored, they are then safe to disperse.

In some species, newborn nymphs crowd beneath their mother's belly or under her wings, much like newly hatched chicks in a henhouse. In a few other species, the mother/offspring relationship is more enduring. Nymphs of the Cuban burrowing cockroach (*Byrsotria fumigata*) remain beneath or near their mother past their second molt—in some instances as long as a month after hatching. Unlike chickens, mother cockroaches do not

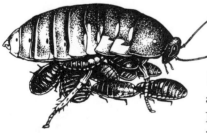

Cape Mountain roach
(*Aptera cingulata*) with babies.

actively round up or herd their young. Researchers believe that these good mothers produce an aggregating pheromone (See Come-hither chemicals, p. 94), that encourages their broods to stick around.

Even more remarkable is the bond between the mother and young of *Perisphaerus semilunatus*, a shiny black cockroach with no common name, a denizen of teak forests in northern Thailand. The nymphs of this species actually cling to their mother's underside, getting a month's worth of free rides through the thickest parts of the forest. Should army ants or other cockroach predators appear, the armor-plated mom acts like a pill bug, deftly rolling herself into a ball to shield all nine of her young.

P. semilunatus nymphs do not have eyes, but gain these as they progress through the various stages of molt. Their long, proboscis-like mouthparts are designed to "plug in" to the small gaps in the mother's belly armor. Blind, nearly helpless hangers-on, the nymphs are exceptionally dependent by cockroach standards. Some cockroach specialists believe that these wimps suckle on glandular secretions from their parent on the go.

Albino and mutant roaches *do* exist, but comprise a *significantly* smaller part of a population than molting nymphs. American cockroaches with "pearl" and "lavender" eyes have been described by Dr. Mary Ross of Virginia Polytechnic Institute, who has also published papers on "odd-bodied" and "balloon-winged" mutants.

Forty-five million from one

How many offspring can one cockroach produce over the course of a lifetime? It depends on who and what you are willing to believe.

The most shocking statistics—and the ones that many cockroach researchers and most pest control practitioners cite most often—are obtained from the German cockroach, one of the most prolific breeders. Over its 150-day life span, one adult female German cockroach can drop as many as eight egg capsules, each filled with as many as forty eggs. That's 3,200 young ones in less than five months.

From this kind of information, creative number crunchers have made mathematically accurate but outlandish projections. The best of these appears in *Ecology and Management of Food-Industry Pests*, a technical bulletin of the U.S. Food and Drug Administration:

> One fertilized female could theoretically produce over 10 million females within one year (about 3.4 generations) and over 10 billion females in just 1.5 years (five generations).

Numbers obtained from actual laboratory studies of German cockroaches are much less dramatic. Starting with ten pairs of adults, Mary Ross reared fifty-one thousand German cockroaches over seven months in 1976. This is far fewer than the USDA's population projections would suggest. Even smaller are the numbers obtained from a lab at Fairleigh Dickinson University, where in one experiment, a male German cockroach was said to have sired 902 young during one short but sweet lifetime.

Both theoretical and laboratory-obtained figures about cockroach populations must be taken with a large grain of salt. That's because neither are based on information from the real world. They assume that the survival rate for cockroach nymphs is 100 percent, not taking into account the effects of predation and disease. They also assume that food supplies are adequate to sustain these thundering herds. And they manage to ignore the possibility of human intervention at some point, or the effects of things such as pesticide bombs on the rapidly building cockroach populations.

Despite these rather large leaps of logic, inflated statistics about cockroach fecundity are still presented as fact in scientific journals and trade magazines. Perhaps the best of these can be credited to Dave Barry. Writing about cockroaches in his home state of Florida, the syndicated columnist observed:

> A single fertilized female Asian cockroach, if all her descendants survive and reproduce, can be responsible in one year for *10 million new cockroaches*. To give you an idea of how many cockroaches that is: if you were to step on five of them per second, 24 hours a day, seven days a week, your shoes would be disgusting.

Worse-case scenario

There's a true story that illustrates how pest cockroach populations can build. Arnold Mallis (in *Handbook of Pest Control*) describes in detail a four-room apartment in Austin, Texas, with an estimated population, prior to treatment, of 50,000 to 100,000 roaches. The majority of these were identified by Mallis as German cockroaches.

"Prior to 1983, we had always visually inspected drawers, cracks, crevices, and cabinets and counted the cockroaches. But residents in the 1980's became less cooperative, and we became more uncomfortable looking through all their belongings. Often we would find automatic weapons, illegal drugs, and other paraphernalia that the residents did not want us to find...The presence of illegal drugs has also changed the types of conditions we see on a daily basis. Although the vast majority of residents are cooperative and want some method of control for what they feel is a serious problem, we have seen children less than one year old abandoned in apartments, child abuse, spouse abuse, and drug abuse. For these reasons there are few researchers who are willing to do field work with cockroaches."

—Philip G. Koehler and Richard S. Patterson, "Cockroaches" in *Insect Potpourri* (1992)

Three weeks after a crew of pest-control specialists treated this apartment with 3.5 quarts of a lethal spray, some four hundred German cockroaches were still infesting the premises. Three weeks after that, the specialists found a population of around a thousand. Approximately six months after the initial spraying, the house was again inspected for pests.

> On approaching the wood shelving over the kitchen sink in order to flush the roaches out with the aerosol bomb, it was observed that …German roaches of all stages were clustered on the walls in great numbers, particularly in the corners where the shelves joined the sides of the open cupboard. Apparently there were so many roaches in the cracks and crevices between and behind the shelving that the roach population overflowed on the walls into the open, which is unusual for German roaches. On directing the aerosol into the cracks and crevices, the German roaches began to emerge immediately in enormous numbers and scurried frantically over the walls and ceilings. Upon lifting the oil cloth of the kitchen table, the edge of the table was found to be encrusted by a great mass of German roaches. These began to fall like rain drops and frantically scatter upon contacting the mist from the aerosol bomb.

Even after half a year of fairly powerful pesticide treatments, Mallis estimated that between fifteen thousand and twenty-five thousand German cockroaches called this crowded kitchen their home.

Movin' on

Left unchecked, a booming cockroach population will eventually exceed the carrying capacity of its indoor environment. When this happens, the excess animals must move out in search of new spaces and food sources, or stay put and suffer the consequences of living in a roach-eat-roach world (described in detail on page 102). Mass migrations are rarely reported, perhaps because these large-scale movements take place at night.

On a dark, drizzly September day in Washington, D.C., entomologist Leland O. Howard witnessed one of these rare departures. "The army issued from the

> "Downtown, on warm nights, American roaches emerge by the thousands from storm drains to find food, moving like a second skin through cobbled alleyways, slipping under back doors of restaurants. En masse, they look like a plague out of some Cinemascope biblical epic."
>
> —Dave Gardetta, *Cucaracha*, from the June 1995 issue of *Los Angeles* magazine

81

rear of an old restaurant fronting upon Pennsylvania Avenue and marched across the muddy street, undeterred by pools of water, ash heaps and other barriers, directly south to the front of the building opposite," he wrote in 1895.

The intended destination of this particular swarm was guarded by a hastily assembled militia of machine shop workers with brooms. "They swept until their arms were tired," continued Howard, adopting a writing style reminiscent of Jack London. Still the men were unable to stem the advancing tide:

> The foreman then directed that a line of hot ashes from the furnace be laid along the brick sidewalk. This proved an effective barricade. The foremost cockroaches burned their antennae and their front legs and the army divided to either side and scurried down into the area ways of adjoining buildings in which they disappeared.

The onslaught of what Howard later identified as thousands of German cockroaches was said to have lasted from two to three hours. "The majority of the individuals composing the army were females carrying egg cases," he later recounted.

Upon learning that there had been no recent house cleaning or insecticide treatments at the restaurant, Howard racked his brains to explain why these females so suddenly decided to move. His best guess? That the urge to migrate arose from their fear that "while the restaurant might support the mothers, there would not be food enough for the coming children." Without any evidence of the cockroach's higher motives, this answer is far from satisfactory.

Cockroach Transportation

HOW A PEST COCKROACH CHOOSES to spend its time (feeding, running, resting, reproducing, and so on) has been closely studied. We know, for instance, that adult male German cockroaches are more active than either females or nymphs. Under normal conditions, these cockroaches' daily rhythms include two phases of high activity. One takes place during the three hours after sunset, the other during the hour before dawn. It is during these two periods that individuals routinely forage for food and water, seek new and better harborage sites, or look for prospective mates.

In indoor environments, German and American cockroaches feed at night. Even when food and water are withheld after dark but made available during the day, the activity patterns of these animals will stay the same. However, as their numbers swell, individuals must become more competitive for food, mates, and harborages. This makes them appear to behave in a more brazen manner, moving to previously unoccupied parts of the home and spending more time in the open—even under the bright light of midday.

Because it's much more difficult to monitor cockroach activities outside of a home or laboratory, detailed activity patterns have not been drafted for the many thousands of nonpest species. When such information is compiled, it's quite possible that our generalizations about a roach's daily life will change.

Fancy footwork

Although cockroaches have six legs, they usually move only three at a time. The first and third legs on one side of the body and the second leg on the other side remain stationary, forming a tripod to support the cockroach's body. Freed from their supporting role, the three other legs move forward. Having taken one step, these legs form a fresh tripod, so the legs that were stationary can now take a step.

This sequence of steps ensures that a cockroach's center of gravity is always within the support area provided by its legs. Thus, the evenly distributed cockroach can stop at any time in the walking pattern without toppling over—something that horses, humans, and many other animals in motion can't do.

A walking cockroach can break into a run simply by picking up the pace of its movements. When brownbanded cockroaches and a few other species are really cruising, they spread their wings and shift their body weight to the rear. With their wings angled upward like a Stealth bomber's, these creatures attain peak speeds by running on their two hind legs.

Workers at the University of California's Berkeley campus have clocked sprinting American cockroaches specimens at speeds of around fifty-nine inches per second. This pencils out to roughly 3.4 miles per hour. It may not seem like much to most people, who can easily walk at more than twice this speed. However, the roach's feat must be considered on a proportional scale.

American cockroaches can cover fifty body lengths per second. That's about ten times the number a human can cover in the same amount of time, and more than eight times the number that a horse can cover. It's nearly three times the relative speed of a cheetah, the fastest animal on land, which can reach a peak speed of forty-five miles per hour.

Considerably smaller than their American relatives, German cockroaches travel at a proportionately slower pace. One of these runners is having a great day if it can go faster than a foot per second.

Roach crossing sign from Atlas Screenprinting

A day at the races

Each spring, thoroughbred racing cockroaches go for the gold (or, in this case, fifty dollars and a trophy shaped like a garbage can) at The All-American Trot. The Kentucky Derby of the six-legged set, this exciting event is hosted by Purdue University's Department of Entomology. Since 1991, students and staff of this department have supplied the custom-built circular track and recruited the racers from their own research stocks.

"These puppies were born to run," claims Arwin Provonsha, curator of insect collections and announcer for the cockroach races. His contestants have names like "Fluttering Antenna," "Hot to Trot," and "Plain Disgusting." They wear racing colors, artfully applied in acrylic paint on each animal's back. Pedigrees ("Seattle Sewer" by "Sewer Sam," out of "Septic System Sally") are displayed on a big board.

Spectators—more than seven thousand in 1995—are encouraged to place their bets for what Provonsha calls a "two-furshort" (as opposed to a two-furlong) race. He asserts that the event has been sanctioned by the Indiana Roach Racing Commission, and that "betting is permitted under their auspices."

Provonsha's pedigreed racers are kept in the dark until the starter's gun. Then they're turned loose into the bright light of day, and, if necessary, prodded into action. Sometimes, the cockroaches run in the same direction.

After continued prodding, the roaches head into the far turn and make a dash for the finish line. Occasionally there's a photo finish, with a *Periplaneta* winning the trophy by a labial palp.

Following the All-American Trot, the action shifts to the three-foot-long straight track, scene of the exhibition tractor pulls. For this event, three Madagascan hissing cockroaches lug miniature green-and-yellow John Deere tractors. The first to tow its load across the finish line—a distance of three feet—is the winner.

Both tractor pull and race are staged within the confines of what Purdue entomologists have dubbed Roachill Downs. Designed and constructed by Provonsha, this unusual backdrop features a grandstand full of dead roaches in sunglasses and baseball caps, waving pennants, drinking soda pop, and eating what their outfitter calls "Green Gunk on a Stick." Other dried specimens pose in lawn chairs, go for grasshopper rides, or wait in a long line outside a port-a-potty.

Are you ready for robo-roach?

Extensively studied by anatomists and physiologists, cockroaches are ideal models for engineering experiments. Students and faculty at Massachusetts Institute of Technology, Case Western Reserve University, and the University of Illinois, Urbana-Champaign have devoted substantial energies to the development of robotic roaches.

One such robot, a 22-inch-long, eight-inch-tall aluminum beast called the Biobot is the brainchild of Fred Delcomyn, a University of Illinois neurobiologist who has spent twenty years studying cockroach locomotion in the lab. Funded by an initial $400,000 grant from the National Science Foundation, the Biobot project may give rise to a new generation of agile and adaptive robots, capable of negotiating dangerous or inaccessible terrain. Obviously, these capabilities would be valuable in space exploration or warfare. However, according to Delcomyn, robotic roaches could also come in handy

around the house. "Most of the robots that are currently available require that you change the environment significantly—clean everything, take everything off the floor, and make sure nobody's moving," he says. "But if asked, for example, to 'pick up dirt' or 'catch cockroaches,' one of ours would be able to perform its tasks while avoiding any obstacles in its path."

Flight of the bumble roach

Some cockroach species are relatively strong fliers, able to cover respectable distances and reach fairly lofty heights. Tagged tropical specimens have been recaptured in tall trees, more than 325 feet from their sites of release. In Arizona, Florida, and several other southern states, flying cockroaches can be seen at night, hovering around streetlamps and other sources of outdoor illumination, especially when air temperatures exceed eighty-five degrees.

Although they are strong fliers, cockroaches are not particularly good navigators. They frequently collide with people and other impediments to their progress. Such collisions are commonplace in Hawaii, where on warm, breezeless nights, male American and Australian cockroaches often take to the sky.

"People here actually think these cockroaches dive bomb them on purpose," notes Lynn LeBeck, an entomologist on the staff of the University of Hawaii at Manoa. Misinterpreting such bumbling attempts at flight for acts of aggression, Hawaiians have nicknamed this creature the "B-52 cockroach."

Florida's feared flier

A flying species that has attracted more than its rightful share of attention is the Asian cockroach, *Blattella Asahinai*, a half-inch to five-eighths-inch-long native of China, southern India, and Southeast Asia. This light-brown insect was first noticed in the United States during the winter of 1984, when it started swarming around streetlights and floodlamps of Kathleen, Florida. Whenever this flier or its compatriots entered people's homes, it headed straight for television screens and other illuminated surfaces. Drawn to areas of brighter light, it seemed to be following residents as they moved from room to room, turning on lights.

This cockroach's airborne abilities and its attraction to light made pest controllers rethink their original prognosis—that they were dealing with a new strain of German cockroach. It took a panel of international cockroach experts to verify that this was indeed a newcomer to North America. Floridians could now add one more cockroach to their state's already impressive species list.

Like its German cousin, the Asian cockroach is light brown with darker longitudinal stripes on its pronotum. Its wings are a bit longer, extending beyond the tip of the abdomen to conceal the female's protruding ootheca from above. This egg capsule is smaller than the German's, but contains the same number of eggs. Such differences, however, are so subtle that even taxonomic experts have difficulty telling the two species apart. Absolute accuracy can be assured only with a gas chromatograph, a sophisticated laboratory instrument that can isolate the constituents of the cockroaches' cuticular waxes.

Scientists suspect the Asian cockroach reached its new home in Kathleen via the Port of Tampa, where it may have been introduced in containerized shipments from the Far East. Originally believed to be confined to the area around Kathleen, the species was discovered to have extended its range first to Lakeland, some sixty miles further inland, then to Ocala, another eighty miles from the seacoast. After two years, populations had spread over 600 square miles. Since then, Asian cockroaches have invaded at least thirty counties. These prolific animals have infested every citrus grove in the Central Ridge area of Florida—occasionally in densities of 100,000 cockroaches per acre. Here they nibble on leaves and branches of orange and grapefruit trees, threatening the livelihood of area fruit growers.

Even more frightening to Floridians is the possibility that the Asian fliers could successfully mate with pesticide-resistant German cockroaches, already ensconced throughout most parts of the state. Such a union could produce near-invincible hybrids, which would rapidly outstrip regional food resources.

Clinical tests have indicated that the Asian and German cockroaches *can* interbreed, if they so choose. However, with the two species exhibiting such markedly different behaviors, it's unlikely that any pairing would take place without the forced intimacy of a laboratory's breeding tank.

Bathing blattarians

Although highly reliant on moisture, few cockroaches have taken the plunge and become wholly aquatic. Only a handful of swimmers have been identified, all from the subfamily Epilamprinidae.

Strictly speaking, these roaches are amphibious, spending as much time out of streams and pools as in them. By pursuing this dual lifestyle, aquatic species can avail themselves of the best fare from both worlds, savoring such tasty morsels as pond scum and the decaying carcasses of fish. A few have been collected from the reservoirs of bromeliads—tropical plants whose water-filled cups can also contain larger life forms, including tadpoles and adult frogs.

In January 1944, while dip-netting mosquito larvae from a lagoon in Panama, entomologist H. H. Crowell captured a single specimen of what he first thought was a common water bug. However, he soon recognized his catch as a cockroach nymph. He brought this prize specimen home, placing it in a commodious battery jar aquarium decorated with bits of water hyacinth and other floating aquatic plants. The little aquanaut survived for weeks in this artificial Eden, thriving on a diet of plant parts and pablum.

> "These nasty and voracious insects fly out in the evenings, and commit monstrous depredations.... They often fly into persons' faces and bosoms and their legs being armed with sharp spines the pricking excites a sudden horror not easily described."
>
> —author unknown, *Natural History of Insects*, vol. 2 (1830)

Eventually tiring of his pet, Crowell shipped his roach to the National Museum in Washington, D.C. It was later identified as a specimen of *Epilampra abdomennigrum*, a familiar black-bottomed blattarian from southern Mexico, Brazil, and several Caribbean islands.

"During the time the roach was under observation in the aquarium, it was induced several times to submerge, voluntarily, by passing the shadow of one's hand over it or by touching it lightly with the end of a pencil," Crowell told readers of the journal *Entomological News*. "When the roach was quiet under the water, a large bubble of air could be seen trapped beneath the pronotal shield." A dive typically lasted a minute or two, during which time it swam briskly around its tank. Then it rested near the surface, holding on to the dangling roots of the aquatic plants for as long as fifteen minutes. Assured that the coast was clear, the roach would climb out of the water and onto the raft of floating plants.

Why were air bubbles trapped beneath the pronotum of Crowell's swimming cockroach? Naturalist Robert Shelford answered this question after closely observing captive water cockroaches from Malaysian Borneo. Ordinarily, he reasoned, an insect's heavy, chitinous exoskeleton would cause it to sink like a stone. However, air filling the tracheae makes this animal bob at the surface like an innertube.

To swim and dive with ease, insects must balance their buoyancy. For aquatic roaches, this means taking quick, shallow breaths. "If closely watched," Shelford wrote in *A Naturalist in Borneo*, "it will be seen that the abdomen gently moves up and down with a regular action, and that there appear at the submerged thoracic spiracles at regular intervals bubbles of air, which grow in size and then break away to give place to fresh bubbles."

At ease

Cockroaches spend considerably more time resting than they do walking, running, flying, or swimming. In the case of the German cockroach, this activity (or lack thereof) can claim as much as three quarters of the day. All other activities, such as exploring, feeding, courting, copulating, and fighting, are crammed into a few hours after dark and again one or two hours before dawn.

In the laboratory, American cockroaches are seldom in motion for more than a few seconds at a time. This is not due to a lack of stamina: on the lab equivalent of a treadmill, healthy representatives of this species have run for four hours without pause.

How can you tell if a cockroach is taking a nap? Typically, it will hunker down, placing its abdomen and egg capsule on the ground. Its antennae will be pointed forwards, angled slightly upwards, and spread about sixty degrees apart. Resting roaches appear to be comfortable with their heads pointing downward or upward—it makes no difference to them.

Because roaches are most comfortable in snug and secure spaces, it's not uncommon during periods of rest to see a row of antennae protruding from cracks or crevices. According to American cockroach authority William S. Bell, some individuals are content to cram into resting places occupied by other species. Others are repelled by such mixed assemblages and will avoid these places. Go figure.

Cockroach cleanliness

For much of their down time, roaches are busy grooming themselves. "Our native cockroaches are, most of them, outdoor feeders and are exceptionally cleanly," Leland O. Howard boasted in *The Insect Book*. "In fact, any of the domestic cockroaches, if watched, will be seen constantly to make efforts to beautify its person, licking its legs and its antennae in much the same manner in which a cat washes its paws."

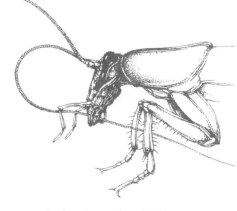

An American cockroach takes time to tidy itself

"Cockroaches clean themselves all the time," agrees Betty Faber, in a recent issue of *Pacific Discovery*, the California Academy of Science's monthly magazine. "You may think a cockroach is filthy, but it thinks you are." A Rutgers University-educated entomologist, Faber now works at the Liberty Science Center, where she enjoys fame and notoriety as the "Roach Lady" of Jersey City, New Jersey.

Before she went public, Faber spent five years as a researcher at the American Museum of Natural History in New York. Here she devoted months to observing the activities of American cockroaches. Her subjects were conveniently located in a greenhouse atop the museum. Spying on these insects at night with an infrared "snooper scope," she took copious notes on the lives of over two thousand individuals. Faber says she became so attuned to the cockroaches' rhythms that she could actually hear their footsteps as they came and went.

Almost all roaches groom their antennae fastidiously. This is the only way they can keep these sense organs in peak shape for receiving scents and vibrations. Thick bristles on the inner set of jaws are for scrubbing the receptor sites of each antenna as these appendages are pulled, much like dental floss, through the cockroach's mouth. Most cockroaches also clean the spines and sensory hairs on their legs, presumably for the same reason. Many insect poisons, especially powdered forms, are designed to adhere to these frequently cleaned surfaces. In this way, the poisons are accidentally ingested by the roach, which, in most instances, would probably leave this foul-tasting stuff alone.

Keeping in touch

Like aphids and scale insects, which instinctively huddle close to each other while feeding on plants, cockroaches are behaviorally driven to stay in touch. However it's not physical contact with other cockroaches, but rather the feel of floors, walls, and ceilings that turns these invertebrates on. The technical term for this behavior is *thigmotaxis*—a predilection for pressure. This strong need to be touched on all sides at once ("a thigmotactic's greatest joy," according to biologist William Beebe) affects nearly every aspect of a blattid's life. Even the flattened, well-oiled exoskeleton of this insect, marvelously suited for wedging into small spaces, has evolved in response to the roach's thigmotactic bent.

All cockroaches are possessed by the hunt for ideal hangouts, more formally known as a harborages. In the wild, suitable harborages include nooks and crannies in peeling tree bark, the moist fissures of limestone caves, or the warm, wet layers of leaf litter that carpet the tropical forest floor. For domestic species, the thigmotactic quest leads to cabinets; gaps between bathroom tiles; pockets behind stoves, refrigerators, and water heaters; and cracks in baseboards, window and door frames, and ornamental trim.

Research conducted by Phil Koehler, Charles Strong, and Richard Patter-

Urban pest cockroaches may enjoy many a free meal, but even they get stuck with the tab now and then. At least one species of mite, a minuscule blighter with a rather long scientific name—*Pimeliaphilus podapolipophagus*—is known to sup on the body fluids of live roaches. As many as 25 of these mites have been counted on the body of a single insect. "A roach so burdened will run about very actively, apparently seeking a means of escape," wrote biologist Frederick Cunliffe, one of the first to focus attention on these haemolymph-sucking parasites. "After about one hour of such heavy parasitism, the roach succumbs, falling over on its back. It thrashes about for some five hours before dying." A bad way to go.

son showed that German cockroaches of various ages occupied different-sized harborages. More than 1,700 laboratory-reared adults and nymphs were placed in containers and given the choice of eight new Plexiglas homes. Each home had a different-sized interior—what the researchers called its "harborage width."

In the home with the smallest harborage width, the ceiling was a sixteenth of an inch from the floor, providing a living space with nearly the thickness of a nickel. This became the favored haunt of one- and two-week-old German cockroach nymphs, who crammed into it with gusto. Adult males headed for the home with the largest harborage width, a spacious half-inch between ceiling and floor. Gravid adult females selected substantially tighter quarters, around three-sixteenths of an inch—about the thickness of three stacked quarters. Three- and four-week-old nymphs checked into harborages in the intermediate sizes.

Shared quarters

This is not to say that in real-life situations, adults and nymphs in various stages of development don't share the same harborages. However, in a reversal of our own way of doing things, the littlest nymphs spend the least amount of time in their homes. Possibly this is by choice, although it's probably a reaction to hostility (See Aggressive Tendencies, p. 17) on the part of adults.

Outside of their harborages, the nymphs huddle shoulder-to-shoulder. For some unknown reason, such aggregating behavior increases the rate of nymphal growth. Because nymphal development is temperature-driven, it could be that the youngsters benefit from such group gropes, which may hold in metabolic heat.

Entomologist J. von Landowski found in the laboratory that heavier adults were obtained when American cockroach nymphs were reared in isolation. He attributed this to an "interference factor"—that is, the frequent collisions between nymphs raised *en masse* did more to slow growth than shortages of food, lack of oxygen, or buildups of feces.

Adult roaches may occasionally abandon one harborage to become a part of a larger group elsewhere. It's also thought that adult males and females

> "Keeping the premises spic and span, whether it is a five-room bungalow or a large macaroni factory is more than half the battle in roach control."
>
> —Arnold Mallis,
> *Handbook of Pest Control*

that have "left the nest" may return to their former harborages to mate, and then go back to their lives elsewhere. In this way, one female can add new life to previously unoccupied harborages—single-handedly causing population explosions in rooms that had been roach-free.

Come-hither chemicals

Pheromones play a crucial part in cockroach cohabitation. An attractant in the feces of German cockroaches and several other blattid species actually draws nymphs and adults together. The highly volatile chemical compounds that make up this pheromone are held in by a thin membrane, secreted around each pellet as it passes through the cockroach's excretory system. The potent attractant (as-yet unnamed by scientists, so feel free to come up with your own name) is slowly metered out over a period of at least one year.

Studies have established that certain large molecules in cockroach saliva can override the effects of aggregation pheromones. In sufficient quantities, these molecules (known as dispersal pheromones) can actually induce others cockroaches to leave overcrowded harborages. The adaptive advantages of these pheromones are also unclear. It's thought that cockroaches may employ these chemicals as emergency measures, only when fast-growing populations are in danger of exhausting food and water supplies. Others have suggested that cockroaches produce these pheromones to warn of impending danger from predators. This second theory is undermined by actual observation: cockroaches coming in contact with the scented saliva back away slowly— not quickly, as one might expect of a creature that had just been tipped off to a threat.

Zapped

Roach motels and other control devices often capitalize on their victims' thigmotactic instincts. One of the more elaborate of these is the Zapper, the electronic brainchild of Greg Jeffreys, Australia's 1991 Inventor of the Year. Made of molded gray plastic and shiny stainless steel, this streamlined appliance resembles the flying saucer from the 1950s thriller *The Day the Earth*

During the Vietnam War, the U.S. Department of Defense was said to have developed a novel approach for detecting farmers doubling as communist guerrillas. Suspected meeting places for Viet Cong guerrillas would be sprinkled with synthetic female cockroach pheromones. Then South Vietnamese police would round up the inhabitants of nearby villages, making them walk slowly past cages containing male cockroaches. The cockroaches would react to any villagers smeared with the synthetic female scent.

Stood Still. The self-proclaimed "Cockroach Dundee from Down Under," Jeffreys says his Zapper is the result of seven years of extensive research. In controlled laboratory tests, it caught 55 percent of all test roaches; in another it reduced populations by 90 percent in six days.

Attracted by the scent of a special pelleted bait, cockroaches enter a narrow chamber on the way to the imagined feast. In so doing, they step onto an electrified metal plate. Should one of these animals skim the chamber's upper surface with their abdomen, wings, or pronotum, it will complete a circuit and receive a fatal jolt of six thousand volts. "Whammo—fried roaches," Jeffreys is quoted as saying in a news release promoting his device.

Frying the cockroaches, Jeffreys maintains, is more effective than poisons because it kills any egg capsules carried by females. However, others have refuted this claim, showing that zapped oothecae could still be incubated successfully.

Harborage horrors

In the autumn of 1983, a middle-aged woman checked into the emergency room at Charity Hospital in New Orleans, complaining of ear discomfort. Help was summoned. Resident physician Kevin O'Toole peered into the woman's ear canal, where he discovered the problem: an adult American cockroach, comfortably ensconced in this warm and snug space.

Initially, O'Toole didn't think much of his discovery. "At Charity Hospital where I was participating in a trauma training program," he told *Omni Magazine*, "we found that problem almost everyday, particularly in people from lower socioeconomic groups."

It wasn't until he'd found a second adult American cockroach, living in the woman's other ear canal, that O'Toole grasped the significance of this event. "We recognized immediately that fate had granted us the opportunity for an elegant comparative therapeutic trial," O'Toole and his associates, P. M. Paris and R. D. Stewart, later reported in the *New England Journal of Medicine*.

The emergency-room team put mineral oil in one ear canal and sprayed the second with 2-percent solution of the anesthetic lidocaine. The roach in mineral oil "succumbed after a valiant but futile struggle, but its removal required much dexterity on the part of the house officer." The other animal, sprayed with lidocaine, leapt from the opposite ear "at a convulsive rate of speed," making its way across the floor in a desperate attempt to escape. "A fleet-footed intern promptly applied an equally time-tested remedy," dispatching the insect with what O'Toole, Paris, and Stuart laconically described as "the simple crush method."

Gastronomy

WHAT DO COCKROACHES EAT? Well, what've you got?

"Bark, leaves, the pith of living cycads, paper, woolen clothes, sugar, cheese, bread, blacking, oil, lemons, ink, flesh, fish, leather," wrote the British scholars I. A. C. Miall and Alfred Denny in their 1886 edition of *The Structure and Life History of the Cockroach*. Add to this list some odd staples like bat guano, rotting tree stumps, crepe de Chine, and their own shed skins, and you've barely begun to compile the cockroaches' complete bill of fare.

"They also eat the corks of bottled wine, cider, and porter, causing the liquid to escape," noted one English colonist identified only as Sells in Glenn W. Herrick's *Insects Injurious to the Household and Annoying to Man*. Sells dubbed these spoilers of spirits "the most annoying of the insect tribes in Jamaica."

"A croton bug of the usual inquisitive turn of mind inhabited my office desk, and as soon as I laid down my cigar upon the edge of the drawer, the little fellow invariably came out of his hiding place and worked vigorously at the moist end," wrote Leland O. Howard, chief of the United States Department of Agriculture's entomology division in 1910. Initially, Howard's deskmate was interested only in the cheroot's moisture content. But eventually, the cockroach became addicted to the tobacco itself. "It may be worth mentioning that it seemed to have no appreciable effect on its health," Howard declared, no doubt defending his own reliance on the pernicious weed.

Is there anything that roaches won't eat? "Cucumber...disagrees with them horribly," offered Miall and Denny. "They have a great dislike to castor oil, which is accordingly rubbed over boots, shoes and other leather articles to protect them," suggested Sells. Certain toxins in the leaves and fruits of the tomato (known in colonial Jamaica as the "cockroach apple") and a few other plants seem to put these animals off their feed. In the United States and Europe, chefs still barricade their work stations with cucumber peelings to ward off six-legged snackers.

According to Dr. Louis Roth, German cockroaches were fed samples of pulverized rock collected from the surface of the moon. This way, NASA scientists hoped to determine if the samples were in any way toxic or capable of infecting organisms on earth. "No evidence was found for the presence of toxins, or infectious pathogens," Roth wrote in his unpublished autobiography, "nor was there any detectable histological injury to the gut of the insects."

In females of the Costa Rican genus *Xestoblatta*, food preferences and feeding behaviors change as these animals enter into different phases of their ovarian cycles. Early in this cycle, females keep to a low-nitrogen, high-lipid diet, dining almost exclusively on the shed bark of *Inga coruscans*, a leguminous tree native to the region. Shortly before they lay their eggs, they shift to more protein-rich fare, then begin a binge on carbohydrates. In many species, females get an extra boost after mating, when males offer them urates secreted from special abdominal glands (see page 72).

Junk food junkies

Most roaches favor starches and sugars over fats and proteins—a gustatorial generalization confirmed by entomologist Phil Rau in 1945. Rau baited seven traps with an assortment of foodstuffs, then placed them in a row, six inches apart, on the floor of a cockroach-infested room. By counting the number of cockroaches that came to each trap over an eleven-day period, he obtained a clear picture of their food preferences.

One trap set with a sugary cinnamon bun drew the biggest crowd—

sixty-five adults, the first twenty-nine of which entered his trap on the experiment's third night. Plain white bread also held an appeal, luring forty-four adults of both sexes by the eleventh night. One trap set with bacon was completely ignored, while another with boiled egg received only one visit. "Almost equally unpopular was the celery, for it attracted only two nymphs, and they came in on the eighth night," noted Rau. Clearly the dietary preferences of cockroaches and children aren't all that different.

To rule out the possibility that the cockroaches were being drawn to the traps by the scent of their own kind, the scientist set two more traps baited solely with live cockroaches. No cockroaches came to these traps, confirming Rau's belief that roach odor (see page 50) is not appetizing, even to roaches.

Blattarian book-biters

Booksellers and librarians know about the cockroach's fondness for the vegetable fibers and animal-based glues of the old-fashioned book bindery. Roaches in rare-book shops are also enamored with gold leaf and its adhesive backing—causing books to lose their shiny titles a few letters at a time. All books, but especially thick, old volumes, can absorb and store water from humid air, becoming desert oases for indoor cockroaches during dry spells.

"The Secretary's files are being seriously injured by the ravages of insects or vermin," the United States Treasury Department's acting secretary Hugh S. Thompson wrote in 1888. "The safety of these files before referred to is of very serious importance to the officers of the Department," wrote E. B. Youwmans, the Department's chief clerk. As clerk, Youwmans explained, he was "held responsible for them all (whether eaten or otherwise)." It was his job "to produce any record called for." Obviously overwrought, he argued that "The law does not recognize the agency of insects in this regard."

An inspection revealed at least two species of cockroaches at work. The basement was crawling with American cockroaches, while a healthy population of German cockroaches were ruling a few of the rooms upstairs. The Americans had nibbled on the covers and backs of nearly half the cloth-bound copies of *Senate Report Upon Methods of Business in the Executive Departments*,

> The agile bookworm eats, conceal'd from sight, Also the prowling mouse abhors the light, But be assur'd that Philobiblos knows, The hellish Cockroach is the chief of foes.
>
> —Jared Bean, *Almanac for the Year 1774*

99

which had been neatly stacked in a corner of the basement. "Specimens of excrement found in the shelves near these books no doubt belong to *P. americana*, and the places eaten have similar excrementitious spots upon them," wrote C. V. Riley, called in to study the problem.

Aided by a few smaller species, the German cockroaches in the Treasury's upstairs were munching on the backs of a number of small paperbound reports, evidently attracted by the paste. "Therefore as a remedy for the future, it would seem advisable to use a poisoned paste in the binding of the Government publications," concluded Riley.

A taste for fine art

C. V. Riley's predecessor, Townsend Glover, discovered that cockroaches enjoyed art just as much as literature: "They made a raid on a box of water colors, where they devoured the cakes of paint, vermillion, cobalt and umber alike; and the only vestige left were the excrements in the form of small pellets of various colors in the bottom of the box," he wrote in 1875.

Deprived of art supplies or good books, cockroaches will turn their attentions to architecture. Wallpaper paste is always a favorite, as is the fiber in paper-based insulation and the plaster applied to wood laths in the walls of older homes.

Invasion of the eyelash eaters

"In the house where we were staying [on the upper Paraguay River of Brazil] there were nearly a dozen children, and every one of them had their eyelashes more or less eaten off by cockroaches," wrote Herbert H. Smith in a letter to Leland Howard, published in 1902. "The eyelashes were bitten off irregularly, in some places quite close to the lid. Like most Brazilians, these children had very long, black eyelashes, and their appearance thus defaced was odd enough."

According to Smith, eyelash-eating was found to have occurred most often among the young children, who "are heavy sleepers and do not disturb the insects at work." Smith and his wife routinely brushed cockroaches from their faces at night, "but thought nothing more of the matter." This reinforces the observations of William Catesby during the 1740s. "It is at night they commit their depredations, and bite people in their beds, especially children's fingers that are greasy," the English colonist in Carolina wrote.

Entomologist Phil Rau confessed that he found the stories of cockroaches and sleeping children "incredulous," until one night: "I was awakened by a tickling sensation on my face, only to find upon opening my eyes, a pair of long cockroach antennae playing delicately for sense impressions while the cockroach's extended mouthparts were imbibing moist nutriment from my nostrils." Rau's up close and personal observations would indicate that the alleged eyelash eaters are really attracted to the minerals and moisture from tear ducts, not the small hairs lining them.

> *"I was awakened by a tickling sensation on my face, only to find upon opening my eyes, a pair of long cockroach antennae playing delicately for sense impressions while the cockroach's extended mouthparts were imbibing moist nutriment from my nostrils."*
>
> —PHIL RAU, Entomologist

An appetite for insects

Cockroaches are also good hunters, chasing down and doing in insects smaller than themselves. An English correspondent in Calcutta in 1910, N. Annandale, described how American cockroaches cleared a dining room of flying termites that had blown in through the windows during a heavy rainstorm. Seizing the termites with their mandibles, the cockroaches feasted where they stood or took their prey to other rooms. In either case, they ate everything but the wings. More recent eyewitnesses have reported cockroaches feeding on moth eggs, mosquitoes, sand flies, larval wasps, and a four-inch-long centipede.

The notion that cockroaches are slayers and eaters of bedbugs may have originated with sailors of the 1700s and 1800s. These salts held fast to their faith that shipboard cockroaches were saving their skins from the bites of minute reddish-brown blood-sucking bugs. The sailors shared this idea with anyone they met on land. In Africa during the early 1900s, locals were said to have asked sailors for a cockroach or two for hunting their villages' bedbugs.

In the 1930s, J. S. Purdy even went so far as to reintroduce cockroaches into a household that had just rid itself of them. It was his off-base theory that the roaches would eat the bedbugs that had recently become established, then they could be eliminated for a second time. Unfortunately, all this faith in the cockroach was completely unwarranted. Under the scrutiny of science, cockroaches have proven themselves extremely uninterested in bedbugs. Two Researchers, C. G. Johnson and Kenneth Mellanby, ran a series of trials in the late 1930s, placing starved adult and juvenile German cockroaches in the Petri dishes with well-fed bedbugs. The bedbugs emerged astonishingly unscathed after nineteen days; only six of thirty-seven showed signs of having been gnawed on by roaches.

In a second trial, two adult and ten German cockroach nymphs were so indifferent to twelve adult bedbugs that the research team nearly succumbed to boredom. "All the bugs were recovered alive after a week and the experiment was discontinued," Johnson and Mellanby yawned.

> "They have no objection to licking the blood of a near relative, and once they have tasted it, they cannot stop until they have gobbled up the victim alive."
>
> —Karl von Frisch, *Ten Little Housemates* (1960)

Cockroach cannibals

Cannibalism is common among cockroaches, particularly in laboratories, where insects are typically reared in large colonies with crowded conditions. Within these artificial insect societies, life is stressful for young American cockroach nymphs, who appear to be more cannibalistic than their parents. In the cramped world of the captive German cockroach, the same conditions have different effects on the nymphs, who begin to consume their own kind fairly late in life—after their fourth molt.

Although the concept of cannibalism goes against our own grain, it gives cockroaches an adaptive advantage. It enables them to adjust population densities to fit the quantities of available food. Cannibalism also confers all of the nutrients from an entire population to a small number of survivors, giving them the strength to develop or reproduce more rapidly, and allowing them the chance to rebuild their numbers at a later date. It may also help maintain the vigor of the species by removing any weak and diseased members from the breeding pool. In laboratory colonies, such practices appear to be stimulated by the presence of injured nymphs—suggesting that cannibalism may also serve as a sanitary measure, eliminating some of the mess caused by overcrowding.

In several species, females are clearly more predisposed than males to eat their neighbors. Within dense colonies of American cockroaches, females will regularly ravage their own egg capsules, even when other food is available. Often these insects eat only the keel of the ootheca, opening a crack through which any moisture escapes, causing the eggs inside to dry out before they can hatch. By biting a bigger hole in its side, the females can also reach in and devour the capsule's contents.

One particularly bloodthirsty female German cockroach, according to J. T. Griffiths and O. E. Tauber, attacked every new male that was dropped into her container. After watching this feisty female do battle with a half dozen prospective mates, the two researchers concluded that "older males were more capable of defending themselves against attacks of the cannibalistic females"— a statement that practically begs for its own bumper sticker.

Starvation study

As awe-inspiring as the list of things that roaches can eat is the length of time that they can survive without any food at all. To establish how long roaches could last between meals, researchers took representatives from eleven species, and, after sustaining them on ample supplies of dog food and water for about two weeks, put two thirds of them on severely restricted diets. Half of these study specimens were given food but no water, while the other half received water but no food. The remaining roaches—the controls in this experiment—had access to both forms of refreshment.

The results, published in 1957 and poetically titled *The Longevity of Starved Cockroaches,* proved that almost all of the species tested could survive for a month or more on water alone. Adult female American cockroaches with access to water could live a full forty-two days—almost half of their natural life spans. Individuals whose water was also withheld demonstrated an equally remarkable ability to hang on in the face of adversity: all but the smallest of species could survive two or three weeks on a total starvation diet.

Data from this rather cold-hearted experiment showed that cockroaches

> "Our beds swarmed with cockroaches, which ran over our faces and hands, or fell from the ceiling. These disagreeable animals are as common here as in Brazil; they gnaw everything, and, being quite soft, are crushed by the slightest motion."
>
> —from *Early Western Travels: 1745-1846* by Reuben Gold Thwaites (1966)

could easily adjust to the rigors of long-distance travel. Accidentally sealed, for instance, in a wooden packing case with severely limited food resources (in this case, the glue that helped hold the crate's slats in place), these creatures could remain fairly comfortable for weeks, enabling them to be successfully shipped almost anywhere in the world. Remember this next Christmas.

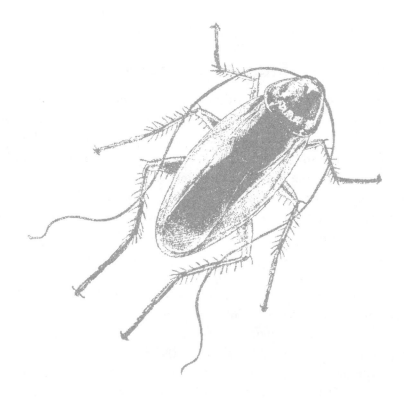

Wasps, Cats, and Other Perils

A LITTLE EXTRA MATERNAL CARE couldn't hurt baby cockroaches (or adults for that matter), considering the many perils that await them. To live to adulthood, cockroaches in the wild must evade a roster of natural enemies—a long list that includes fish, reptiles, amphibians, birds, mammals, other arthropods, and, occasionally, their own kind.

New York naturalist William Beebe identified four species of tropical fish from British Guiana that eat cockroaches—a stingray, a freshwater catfish, and two kinds of characins. While employed at a pet store in Chicago, I watched archer fish spit droplets of water at German cockroaches, knocking them into their tank. Phil Rau wrote about Oriental cockroaches being sold as bait to bluegill fishermen in Tennessee.

Anolis lizards (a.k.a. New World chameleons) relish the soft bodies of immature Surinam cockroaches, according to P. J. Darlington, Jr., who kept these sit-and-wait predators as pets in Cuba during the 1930s. "This insect is probably a staple food of [this lizard] in the wild state," he wrote. "On several occasions, when I have been collecting insects by washing out trash on the banks of streams, wild *Anolis* have come down to the bank and caught immature roaches which I had allowed to escape." F. J. Simmons counted thirty-one roaches from the stomachs of forty-six Anolis lizards in Bermuda.

Cockroaches are frequently taken from the stomachs of frogs, and toads are supposedly even more notorious arthropod gobblers. One of these cold-

blooded hunters is able to "clear your room of cockroaches overnight, just as he will your garden of the vilest of your insect foes," wrote a toad enthusiast in 1889.

On the subject of gobbling, it should be mentioned that wild turkeys, too, have no qualms about downing a few blattarians with their meals. These birds will pick at any food item that attracts their attention. One turkey hen, killed in Virginia in 1939, was found to have consumed an assortment of grasses; twelve acorns; seeds of the buttercup, greenbrier, tall meadow rue, violet, sedge; and a smorgasbord of insect species, including several crane flies, a few stink bugs, a couple of click beetles, some weevils, and "eight or more cockroaches," according to the author of *Complete Book of the Wild Turkey*. As they say, "Roaches never get justice when a turkey is the judge."

Munched by mammals

A rather graphic portrait of wild-caught roaches as mammalian cuisine is provided in an 1829 issue of *Magazine of Natural History*. In this piece, P. Neill, Esq. described the travails of transporting a marmoset monkey to Scotland from Brazil by ship:

> As long as the fruit which he had on board lasted, it would eat nothing else; but when these failed, we soon discovered a most agreeable substitute, which it appeared to relish above everything. By chance we observed it devouring a large cockroach which it had caught, running along the deck of the vessel; and from this time to nearly the end of the voyage, a space of four or five weeks, it fed almost exclusively on these insects, and contributed most effectually to rid the vessel of them.

Neill noted that the monkey could easily swallow a score of the largest insects, plus "a very great number of the smaller ones," three or four times over the course of the day. "The small cockroaches he eat [sic] without such fastidious nicety," observed the attentive biologist. Obviously impressed by the primate's table manners, he continued:

> When he had got hold of one of the large cockroaches, he held it in his fore paws, and then invariably nipped the head off first; he then

pulled out the viscera and cast them aside, and devoured the rest of the body, rejecting the dry elytra and wings, and also the legs of the insect, which are covered with short stiff bristles.

Lemurs, tarsiers, ring-tailed cats, opossums, bats, ocelots, and armadillos are all roach eaters. Specimens of *Epilampra wheeleri*, *Eurycotis improcera*, *Panchlora nivea*, *Pycnoscelus urinamensis*, *Ischnoptera rufa*, *Periplaneta americana*, and *Periplaneta australasiae* have been obtained from stomachs of mongooses in Puerto Rico, Saint Croix, East Africa, and Hawaii.

Arthropod opponents

A wild cockroach's deadliest enemies are usually the same size or smaller than them. Among the worst are army ants, which march en masse, sweeping up and devouring any game in their path, regardless of its size. One of the many *Marvels of Insect Life* in the 1915 book by the same name is the housecleaning service rendered free of charge for the home of a Mrs. Carmichael, proprietress of the Laurel Hill estate on the island of Trinidad. When a horde of these pernicious predators traipsed through her home one spring day, Carmichael followed them from room to room, opening linen chests and storage spaces. The ants quickly consumed every cockroach in the place, and ridded the entire premises of rats and mice as well.

A nocturnal hunter, the vaejovid scorpion is the scourge of certain desert-dwelling cockroaches. This small, venomous arthropod emerges at dusk to prowl the California desert, searching the arid foot of Mount San Jacinto for its favorite prey, the sand cockroach *Arenivaga investigata*. Upon arriving at its favorite stalking ground, the scorpion freezes. Its keen senses and motionless stance allow it to detect the slightest vibrations beneath the sand.

As it tunnels through the first half inch of the fine desert soil, the sand cockroach is an easy mark. The scorpion maneuvers itself directly over its meal and thrusts its pedipalps straight down, grasping the hapless victim and injecting it with venom. Like the marmoset, the scorpion chows down on its prey head first.

In the mid 1960s, the General Motors corporation tested a number of schemes intended to electrocute the cockroaches infesting New York City buses. In one experiment, they installed grids with electrified strips near the buses' ankle-high heating ducts. After a month of field testing, a checkup revealed only two or three dead roaches per bus. "It probably cost $1,000 a cockroach to kill them," according to John J. Courtney, assistant general superintendent of the Manhattan and Bronx Authority, in the November 20, 1978 issue of the *New York Times*. The *Times* article also described the Manhattan Transit Authority's was on German cockroaches, which entailed fumigating its entire fleet of 4,500 buses every other weekend with insecticide bombs.

WASPS ARE THE WORST

It's hard not to feel sorry for cockroaches when certain predatory wasps get their claws on them. One of these, the bright-red and blue-green jewel wasp of central and East Africa, ambushes American and Australian cockroaches from the side, swiftly jabbing with its stinger at the cockroach's first leg. This makes the roach turn and tumble in an effort to escape.

No such luck: after wrestling with its prey, the wasp gains the upper hand, partially paralyzing it with the powerful poison in its stinger. A second, more precisely aimed sting is delivered to the roach's now-unprotected head.

The paralyzed blattarian can now only watch as the wasp takes a break, meticulously preening itself for as long as half an hour. Then the wasp amputates one or both antennae, sucking up the fluid that oozes from the fresh wounds.

Holding the cockroach by its antennal stumps, the wasp drags its catch to a nearby hideout. Here it deposits a single white egg inside its victim. The wasp then seals the still-conscious insect in a tube-shaped cavity that will serve as a nest. In two days, the three-sixteenths-inch-long egg hatches. The larval wasp finds the cockroach good to the last drop: after imbibing the body fluid, it devours what's left of its immobilized host.

You can see the entire gruesome affair—from the first sting to the last larval snack—at the Insect House of the Artis Zoo in Amsterdam. At this facility, jewel wasps are not only encouraged to feed on the roaches in their display, but have also been released in the zoo's small mammal house in an effort to control an infestation of American cockroaches without using chemicals. A similar wasp exhibit is also an attraction at the London Zoo.

Many other wasp species have discovered that roach egg capsules make superb vessels for rearing their own young. These animals, called parasitoid wasps, actively seek the capsules of specific cockroach species. One member of this wasp family ravages brownbanded cockroach egg capsules exclusively. When their three to fifteen eggs hatch—about a week before the baby brownbanded cockroaches are due—the young wasps devour the capsule's rightful occupants. After this species' accidental introduction in Hawaii, nearly 100 percent of all brownbanded egg capsules in some areas had been parasitized.

Biologists in Kobe, Japan, fear that their city's 1995 earthquake may have destroyed prime parasitoid wasp habitats, setting the stage for an unprecedented cockroach population explosion.

Parasitoid wasps of the family Evaniidae lay single eggs in the capsules of every important pest species of roach, with the possible exception of the brownbanded. The larvae of these wasps feed on one egg after another until all are destroyed, making them quite popular with people bugged by roaches. Indeed, the scientific name for one of these, *Blatticida pulchra,* a pint-sized member of the wasp family *Encrytidae,* is translated as "beautiful cockroach-killer." The effectiveness of the Evaniidae has made them sought-after cockroach controllers in home environments.

Indoor enemies

Predators may be less abundant in the world of the more familiar domestic cockroachs. Nonetheless, those predators that *do* exist can be just as lethal. Like the native Jamaicans of previous centuries, who encouraged large cockroach-eating spiders to share their thatched huts with them, we should recognize the value of these and other insectivores in keeping our domiciles cockroach-free.

All spiders are predators, but few are specialized enough to eat any one particular kind of prey. In our homes, these animals dine on a smorgasbord of moisture-loving pests—roaches, bedbugs, carpet beetles, earwigs, houseflies, and, if the opportunity presents itself, other spiders. The eight-legged heroine of E. B. White's classic children's book *Charlotte's Web* was probably right when she proclaimed "if I didn't catch bugs and eat them, the bugs would increase and multiply and get so numerous that they'd destroy the earth, wipe out everything."

The hobo spider (a name that alludes to its principal means of dispersal) was introduced to America from Europe—just like the cockroaches it pursues. After crossing the Atlantic Ocean, this spider made its way west by riding the rails, slinging its web in hobo camps along the way. By 1930, it had established itself throughout the United States.

Its web has all the architectural elegance of a mobile home. Aesthetically pleasing or not, this structure serves as an effective cockroach trap. Hung like a hammock in any out-of-the-way part of the basement, its outer edge curves upward while the rest slants down, forming a funnel that empties into

> "Speed and wariness would seem to be the principal defensive assets of cockroaches where spiders are concerned. Any scent they emit is not obnoxious to spiders."
>
> —William Snyder Bristowe, *The Comity of Spiders* (1939)

a tubelike lair. The surface of this funnel is covered with tiny trip wires, designed to befuddle any blattid that stumbles across it. Whenever opportunity knocks, the spider is quick to respond: it races out of the funnel, sweeps its prey off its feet, and then scrambles adroitly back to its lair for a leisurely cockroach feed.

STALKED BY CENTIPEDES

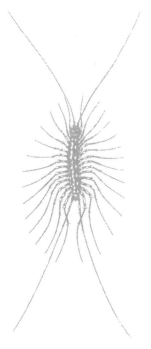

Centipedes are grayish-yellow inch-long creatures, not insects at all but members of the Class Chilopoda. The house centipede has fifteen pairs of legs, each encircled with alternating dark and white bands. Young are born with only seven of these striped stems, but the number increases with every molt.

The last pair of a house centipede's legs are much longer than the rest. They are specially adapted for lassoing and holding insects, spiders, and other small prey. Once captured, these victims are quickly dispatched with one bite from the centipede's "jaws"—in actuality, its modified front legs. These biting legs are connected to glands that secrete a highly effective insect poison.

Originally from Europe and introduced to Mexico, the house centipede is now found throughout the United States. It lives indoors and out, but being a lover of darkness and moisture, it appears to prefer cellars, dank closets, and bathrooms. Like other members of its family, the house centipede is most active at night.

To test this invertebrate's mettle, a University of Maryland entomologist once placed a house centipede in a dish with a female German cockroach and her egg capsule, which happened to be hatching. "No sooner were the young roaches running about," he wrote, "than the centipede began a feast which ended only when the last of the brood had been devoured."

> The mother roach was not at the time molested, but next morning she lay dead on her back, her head severed and dragged some distance from the body, which was sucked dry of its juices—mute evidence of the tragedy that had befallen sometime in the night, probably when the pangs of returning hunger stirred the centipede to renewed activity.

You think *that's* something, wrote Roth and Willis in *The Biotic Associations of Cockroaches*. "Our specimen caught a small American cockroach nymph that we placed in a jar. Before it had finished its meal, it had caught and held two other nymphs with its legs while it continued to feed on the first." Consider this the next time you chase one of these valuable arthropods from your roach-infested home.

MUNCHED BY MICE

House mice are universally disliked, ranked (with rats) a few points below cockroaches on the gross-animal scale (see page 55). However, many individuals in this species may occasionally redeem themselves in our eyes. For quietly at night, when no one is looking, these small mammals catch and eat any cockroaches that encroach on their turf.

To determine how readily these rodents took to cockroaches, mammalogist Mark Wourms captured thirty-one house mice in buildings throughout Boston University and Cambridge, Massachusetts. In a move reminiscent of tossing Christians to the lions, he then released German cockroaches, five at a time, in the same enclosures as these wild-caught mini-beasts.

"Typically, a house mouse approached a cockroach to within 1 cm, partially closed both eyes, and sniffed it," Wourms observed in *Journal of Mammology*. The mouse followed this initial investigation by striking out with its front paws. This caused the roach to run "rapidly and erratically"—a behavior that elicited more rapid pawing and lunging bites from the mouse. After several seconds of such mouse handling, the roach either escaped or was taken hostage. After the prey was immobilized, its head was removed and its elytra, legs, and wings were nibbled off and dropped.

COCKROACH-EATING CATS

The contributions of the domestic cat should not be overlooked as they are known to hunt roaches, for both food and sport. I am aware of only three formal studies that have explored this relationship, and these all involved cats that were cruelly fed cockroaches infected with internal parasites. However, several knowledgeable cat owners have volunteered information that indicates a noticeable decline in blattarian numbers after acquiring kitties. Others have

> She thinks that the cockroaches just need employment
> To prevent them from idle and wanton destroyment.
> So she's formed from that lot of disorderly louts,
> A troop of well-disciplined helpful boyscouts,
> With a purpose in life and a good deed to do—
> And she's even created a Beetle Tattoo.
>
> —T.S. Eliot,
> *The Old Gumbie Cat,*
> from *Old Possum's Book of Practical Cats*

informed me that, should a cat choose to play with its food, the hunt can become quite contrived. Only after a good half hour of catching, releasing, and batting its prey will the cat deliver the coup de grace.

"Mockroaches"

A good way to avoid being eaten is to look like something inedible. This strategy is successfully employed by several roach species that have evolved to resemble other, less-tasty insects. In the Philippines, several brightly colored but foul-tasting ladybird beetles and leafbeetles are impersonated by cockroaches of the genus *Prosoplecta*. The hind wings of these vibrant-hued look-alikes are rolled up and folded, giving their owners a short, rounded shape while at rest. To further pass as beetles, these roaches must hold their slightly longer antennae motionless. The overall effect is so convincing that these insects can cruise their island habitats with impunity during the day. In one species, the male even mimics the color morph of one leaf beetle, while the female mimics another.

Just as astonishing are the disguises worn by members of the Neotropical cockroach *Schultesia lampyridiformis*, whose species name, literally translated as "fireflylike shape," just about says it all. This jaunty fellow's disguise is so convincing that it can scavenge for food in orioles' nests without becoming bird food.

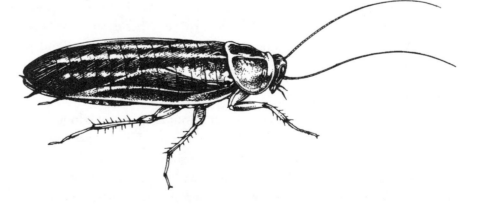

Schultesia lampyridiformis, a neotropical copycat.

Fighting back

Don't think for a moment that only those cockroaches that can hiss loudly or pass as other bugs will escape a predaceous pounce. Sharp spines on the legs of many species are effective defensive weapons, especially against small insectivorous animals. Other roaches have even better deterrents up their sleeves. The Florida woods cockroach (*Eurycotis floridana*) is a wingless species primarily confined to central and southern Florida and the West Indies, which takes up residence in dead stumps, logs, cavities in limestone rock, and piles of firewood. When startled, this roach releases a greasy, foul-smelling mist as a defense mechanism. Produced by glands in the abdomen, this spray contains the aldehyde 2-hexenal—a complex chemical compound that is mildly irritating to human skin. When not actively spraying, however, it possesses an unmistakable scent of maraschino cherries. This pleasing aroma has done little to change this animal's common name, the stinking cockroach.

A rapid retreat

Astonishingly fleet of foot, few animals are more adept at beating a hasty retreat. Even more astounding is the cockroach's split-second response to a perceived threat, which sets the alert insect's legs in motion within a half-second. Students of the American cockroach have determined that this animal responds to puffs of wind and tactile stimuli by turning away from these forces and running a short distance, thus evading any would-be predator. This escape maneuver was once thought to be a simple reflex.

However, in-depth observation has revealed that the American cockroach in this situation must actually process quite a bit of information (Does the wind puff represent a real threat? Will running across an open area make the roach more vulnerable than if it stood still?) before it chooses to bolt. Only after it weighs all of the options will the cockroach make its move—almost always the right one, as anybody who's ever tried to swat a roach on the run will attest.

When Humans and Cockroaches Meet

Cockroaches in Human Culture

A Cockroach masterpiece

COCKROACHES ARE CONSPICUOUSLY absent from the realistic still lives and landscapes of the Dutch masters. They've also been overlooked by the French impressionist painters of the early 1900s and their successors, the expressionists of post–World War I Germany. However, these often-impoverished artists were probably influenced by the many domestic roaches that shared their abodes. Although undetectable in the finished works, these blattids undoubtedly lurked beneath Van Gogh's rumpled bed in Arles and graced the bowls of apples and pears so lovingly painted by Matisse.

The closest thing to a cockroach masterpiece can be credited to Maria Sibylla Merian, a German-born naturalist and watercolor artist. In 1699, at the age of fifty-two, Merian packed up her paints and, along with her younger daughter, sailed to Surinam, at that time a Dutch colony on the northeast coast of South America. Here she studied animals and plants, identifying nearly 100 insect species and painting a series of exquisite nature studies. In 1705, these pieces were reproduced as lithographs in *Metamorphosis Insectorum Surinamensium* (later titled *Dissertation in Insect Generations and Metamorphosis in Surinam*). Plate II of this publication contains images of American and German cockroaches in assorted developmental stages, crawling and hovering around a wild pineapple plant in bloom.

The purpose of her plant- and insect-inspired art, Merian wrote, was "so that gentlemen scholars can see what wonderful works and animals God has created in America." Having admirably achieved this goal, Merian nonetheless died in poverty, twelve years after her paintings' public debut. Many of her original works were purchased by Czar Peter the Great, while others have since been acquired by the British Museum and the Royal Library at Windsor Castle.

Contemporary Cockroach Art

A few distinguished painters and sculptors have incorporated images of cockroaches into their artistic creations. Among the most striking of these are the works of Filipino artist Manuel Ocampo. A few of his larger canvasses include images of roaches, surrounded by symbols of the Spanish colonization and Catholic conversion of Ocampo's native land.

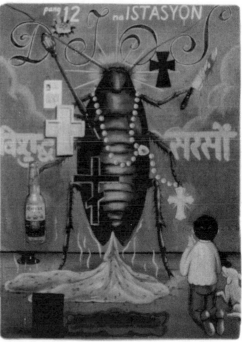

Less political but equally poignant are the 3-D tableaus of New York City artist and jeweler Richard Boscarino. Reproduced as a series of postcards in the 1970s, Boscarino's mini masterpieces feature dried and mounted American cockroaches photographed in unusual poses and outfits, placed in imaginative, highly detailed sets.

One of Boscarino's more popular pieces is a detailed model of a Providence, Rhode Island, diner, complete with a harried cockroach chef and blonde waitress, both bustling to serve a hungry throng of chitinous customers. A youngster, who can barely see over the counter, rounds out this tableau. In "Birthday Party," another of Boscarino's postcard pieces, six festively attired roaches, some in party hats, are seated around a circular table, eating cake, blowing party horns, launching paper airplanes, and spilling soft drinks.

Another important figure in "roachart" (as Boscarino calls it) is Dick Webb, a part-time sculptor and full-time police officer in Ames, Iowa. Webb's work captured the attention of the media in 1982, roughly seven years after he completed his first piece—a gag trophy for a target pistol competition, featuring dead roaches on a replica shooting range. But unlike Boscarino, who purchased his dead specimens from a biological supply company, Webb insisted on collecting his own—at the city jail.

Webb's wonders (clockwise from upper left: a roach
clip; the 9th hole; hot tubbing; *That's Incredible* set.)

"I usually go to the jail just after they spray [every ten days] and I can pick up three or four good ones," he told *The Des Moines Register*. "I throw the small ones back." The freshly killed roaches are more pliable, according to the *Register*, making them easier for Webb to bend into humanlike poses. For extra durability, each subject received two coats of clear polymer.

Word of Webb's unusual trophies spread, and new opportunities to create larger works began to appear as he received commissions from some unlikely sources. Indeed, before long Webb and his wife were fielding queries from a vertitable host of radio and television producers. And to say the least, he rose to the challenge. Prior to an appearance on ABC's *That's Incredible!* the Ames policeman invested more than fifty hours of off-duty time to construct a miniature studio set, with an audience and crew of eighty-eight artfully positioned roaches. Now retired, Webb is content to rest on his laurels, leaving this potentially lucrative cottage endeavor to the next generation of cockroach catchers and posers.

Cockroaches on—and in—TV

For *Post Nature*, an exhibition at the New Langton Arts gallery in San Francisco, Gina Lamb created a piece she named "TV Consumers." This involved importing a colony of lab-reared American cockroaches and placing them in a Plexiglas cage with a dismantled TV set, innards exposed. Inside this environment, Lamb installed a closed-circuit video camera with a zoom lens, so she could project images of the TV's insides, bustling with blattids, on its own twenty-one-inch screen.

"Lamb has fashioned a perfectly self-reflexive and self-sustaining mechanism," noted one critic from *Art News*. "Not only does the monitor feed its own image back to itself but it also nourishes the cockroaches, who feast on the insulation material." To the *Art News* critic, "TV Consumers" offered an apt metaphor "for the current media-saturated state of hyperreality...in which the signs for the real have parasitically consumed their beleaguered host!"

Cockroach comics

Cockroach characters have been woven into the plots of several syndicated comics and a few "subterranean" ones. As frequent visitors to *Outland*, by Berke Breathed, these insects tormented the strip's two main characters, Opus and Bill, and, on one occasion, made lascivious calls (at five dollars a minute) to a "900 Dial-A-Congressman" telephone talk line. And the slow-paced *A Wish for Wings That Work*, an animated half-hour feature set in *Outland*'s surrealist landscape, is enlivened by the brief appearance of a cross-dressing cockroach with coffee-bean breasts.

> **Diner:**
> "Waiter, there's cockroach swimming in my soup."
> **Waiter:**
> "And what would you do in his place, drown yourself?"

Gary Larson's *Far Side* also found high humor in humankind's least-loved creature. In one of these enlightened cartoons, a car full of roaches is shown late at night, pulling up to a three-story brick Roach Hotel. "Dad! Find out if they have cable!" cries a voice from inside the car. An even better one-liner comes from *Real Life Adventures*, by G. Wise and Aldrich. The caption to their drawing of a Roach Spray container reads "For guaranteed results, beat insect with can."

In 1927, George Herriman supplied the comedic pen-and-ink drawings for the original *archy and mehitabel* and for three collections of Archy's verses, published in 1933, 1935, and 1940. Herriman also conceived, wrote, and illustrated the legendary comic strip *Krazy Kat*, which appeared in newspapers across the U.S. for an unprecedented twenty-nine years. *Krazy Kat*'s cast of back-alley characters inspired cartoonist Mark Kausler's design for Malcom the Cockroach, an endearing chap from the 1975 animated film *Coonskin*.

BLATTMAN?

Captain Cockroach is the star of Dave Sim's independent comic book *Cerebus*, serialized since 1977. Sim originally conceived this character—a crime fighter who dresses in a roach costume to disgust his foes—as a one-time parody of Batman. However, The Cockroach proved so popular that the artist established him as a schizophrenic, with multiple personalities that poke fun at other, more famous comic-book heroes.

Domino Chance, Roach Extraordinaire is the invention of comic artist Kevin Lenagh of Minneapolis. Lenagh featured this roach space-fighter from

1982 to 1985 in a series of nine self-published comic books. Chance's adventures are set twelve thousand years in the future, when chaos reigns and human-sized roaches in red, hooded flight suits call the shots.

Assisted by Mel and Frencci River, Stanley White introduced *The Bugged Out Adventures of Ralfy Roach* in 1992, then published three follow-up issues in 1993 and 1994.

COCKROACH CARTOONS

In 1935, a mere six years after Al Jolson's *The Jazz Singer* ushered in the new era of "talkies," roaches appeared in their own screen musical, *The Lady in Red*. This five-minute animated film was directed by Friz Freleng, best remembered for directing the first Bugs Bunny cartoons. It takes place in a closed Mexican cafe, where a bunch of roaches are making themselves at home—bowling with olives, playing tennis with peas, and crowding into the Roach Nite Club, where dancing and musical entertainment is provided by the lovely Senorita Cockroach. Before the Carmen Miranda-esque crooner can conclude her rendition of "The Lady In Red," she is abducted by a parrot, then rescued by a handsome blattarian hero. The end.

The groundwork for Freleng's *Lady* was laid by Paul Terry, the founder of the once-mighty Terrytoon studios. Terry's 1932 cartoon *Cocky Cockroach* entertained matinee-goers with a similar boy-meets-girl, boy-loses-girl (this time to a spider) plot. Both *Lady* and *Cocky* have been the inspiration for dozens of animators, most notably Martin Barry, the French Canadian director of the award-winning *Juke Bar*. A musical comedy that marries puppet animation with live action, this ten-and-a-half-minute short is also full of singing and dancing. It's set inside a vintage juke box, which, at the film's end, becomes a roach trap.

No one has compiled a complete list of cartoons in which cockroaches have appeared, but here are eight of them: the silent *Bug Vaudeville* (1921), *Cocky Cockroach* (1932), *The Lady in Red* (1935), *Bingo Crosbyana* (1936), *The Ugly Cockroach* (1960), the Archy-inspired *shinbone alley* (1971), *The Metamorphosis of Mr. Samsa* (1978), and Don Bluth's *An American Tail* (1986).

This does not include the many made-for-television cockroach cartoons. Notable among these are *The Show-Off Roach* (an episode of the charming children's show *Maya the Bee*) and *Capitol Critters*, a short-lived TV series starring roaches, sewer rats, and other unwanted animal inhabitants of the White House.

A cockroach epic

Twilight of the Cockroaches is the *Gone With the Wind* of insect animation. Written and directed by Hiroaki Yoshida, this ambitious production from Japan utilizes live action and animated footage to chronicle the rise and fall of two domestic cockroach societies. One of these flourishes in the comfortable home of Mr. Saito, a modern bachelor who loves good food and strong drink, the latter making him all but oblivious to his six-legged housemates. In Saito's *goki buri no deru dai dokoro* (a roach-ridden kitchen), there's a party every night. The other group lives across an open field from Mr. Saito's cheery digs. The members of this second society are under constant attack from a fastidious career woman, eager to rid her home of these pests.

© 1987 TYO Productions, Inc./Kitty Films, Inc.

The citizens of the two cockroach kingdoms are brought together after Hans, a wounded roach soldier, staggers through the doorway of the Saito home. Hans warns the party-hearties of the impending genocide. However, the only blattid who believes him is Naomi, a beautiful nymph who is about to be married to a sensitive and poetic young roach named Ichiro. Naturally, while administering to the wounds of Hans, Naomi falls in love with the dark and mysterious warrior.

Things soon turn sour for Naomi's family and friends after Saito brings home his new girlfriend—the woman from across the way! Together, they mount an all-out insecticidal attack on the once-peaceful insect kingdom. At the conclusion of the film, the only survivor is Naomi, now carrying an egg capsule full of fertilized eggs from her brief affair with Hans.

"Everyone takes it a little differently," Yoshida said, when asked to explain his film's significance. "Some Americans think it's about the Jewish people. Koreans think it's about the Korean people. Blacks think it's about blacks," he said. Pressed further, he revealed his true intent:

> These cockroaches of mine are blindly unaware of the situation in their world. They are living a good life without knowing history. My cockroaches are like today's fourth generation of Japanese [since World War II]. They have no idea of danger, and they're not interested in danger; they think the natural thing to do is to live, to eat, to have everything and always have a good time. To suffer from nothing. And that is a dangerous idea, for a cockroach and a Japanese.

Literary cockroaches

Cockroaches waited many millennia for someone to put their inner feelings in print. That someone turned out to be Archy, an ambitious blattid, alleged by newspaperman Don Marquis to have inhabited the offices of the *New York Sun* from 1913 to 1922.

Arriving at his office a little earlier than usual one morning, Marquis happened on what he described as "a giant cockroach" jumping about on the keys of his typewriter.

Scuttle, scuttle, little roach—How you run when I approach; Up above the pantry shelf, Hastening to secrete yourself.
Most adventurous of vermin How I wish I could determine How you spend your hours of ease, Perhaps reclining on the cheese.
Cook has gone, and all is dark—Then the kitchen is your park; In the garbage heap that she leaves Do you browse among the tea leaves? How delightful to suspect All the places you have trekked; Does your long antennae whisk its Gentle tip across the biscuits?
Do you chant your simple tunes Swimming in the baby's prunes? Then, when dawn comes, do you slink Homeward to the kitchen sink?
Timid roach why be so shy? We are brothers, thou and I. In the midnight, like yourself, I explore the pantry shelf.

—Christopher Morley

He did not see us, and we watched him. He would climb painfully upon the framework of the machine and cast himself downward, and his weight and the impact of the blow were just sufficient to operate the machine, one slow letter at a time. He could not work the capital letters, and he had a great deal of difficulty operating the mechanism that shifts the paper so that a fresh line may be started.

Marquis wrote that he "never saw a cockroach work so hard or perspire so freely" as this one. "After about an hour of this frightfully difficult literary labor he fell to the floor exhausted and we saw him creep feebly into a nest of the poems which are always there in profusion." Removing the sheet of paper from his typewriter, Marquis found the following:

> expression is the need of my soul
> i was once a vers libre bard
> but i died and my soul went into the body of a cockroach
> it has given me a new outlook upon life
> i see things from the under side now
> thank you for the apple peelings in the wastepaper basket
> but your paste is getting so stale i can t eat it

These stanzas marked the beginning of Archy's long, literary career. Over the next ten years at the *Sun* and continuing with Marquis' move to the *New York Tribune*, the writings of North America's hardest-working roach poured forth. His brief contributions initiated readers into the rich world of urban insects and kept them abreast of the antics of the alley cat, Mehitabel, for whom Archy carried a torch.

Archy's columns were often whimsical, satirically iconoclastic, and oftentimes bitter. They were also economical. Their short lines covered the maximum number of column inches with the least amount of effort on the part of the author. In addition, relying on the cockroach's correspondence enabled Marquis to avoid using capital letters, apostrophes, and quotation marks—those minor annoyances, according to Marquis' pal E. B. White, "that slow up all men who are hoping their spirits will soar in time to catch the edition."

Published as a compendium in 1940, *the lives and times of archy and mehitabel* contains over 200 of these pieces, alternately addressing various trendy topics ("old doc einstein has/abolished time but they/haven't got the news at/sing sing yet") and touching on universal truths ("procrastination is the/art of keeping/up with yesterday"). Archy's pieces pull no punches, especially where humankind is concerned:

> i do not see why men
> should be so proud
> insects have the more
> ancient lineage
> according to the scientists
> insects were insects
> when man was only
> a burbling whatisit

or

> the cockroach lives
> in peace and plenty
> while the human race
> hustles to support him
> all the social institutions
> of all time have existed
> for the purpose
> of forming a pyramid
> on the apex of which
> perches the cockroach triumphant
> it has taken us a long
> time but we point
> with pride to the achievement

Additional verses by Archy can be found on pages x, 84, and 126 of this book.

Cockroach novels

In *The Cockroaches of Stay More*, by novelist Donald Harington, readers are treated to the views of several cockroach families who live in and around the houses of a rural Ozark community.

Harington takes his time telling *Stay More*'s plethora of interwoven, and often melodramatic, tales. In the process, he provides many fresh perspectives on the subjects of roach biology and misbehavior. Readers learn, for example, that young blattarians are taught by their mothers to "gobble yore food, but puke in solitude," and that, "Above all...it is not nice, ever, to use the word 'cockroach.'"

The personable arthropods of *Stay More*—flirtatious Letitia "Tish" Dingletoon, the Reverend Chidiock Tichborne, the aging, almost deaf bachelor Sam Ingledew, and their many family members and friends—prefer to call themselves "roosterroaches." Much of this book's convoluted plot centers around what the roosterroaches call "Holy House," so-named because its human occupant (a weak-willed, alcoholic writer) shoots holes in its walls, in an attempt to exterminate his insect cohabitants with a pistol.

The Roaches Have No King, by Daniel Evan Weiss, puts a streetwise spin on Harington's story of cockroach country life. Its setting is New York City—specifically the apartment of Ira Fishblatt, a liberal and slightly less-than-hygienic attorney. The narrator, Numbers, spent his first days of life nibbling away at the Old Testament on Ira's bookshelf. Weaned from books to kitchen comestibles, Numbers must avert the life-threatening changes that will occur as part of a proposed kitchen remodeling. To save the day for his species, he must destroy the relationship between Fishblatt and his fastidious new girlfriend (Shades of the film *Twilight of the Cockroaches*), Ruth Grubstein.

Roaches was originally published in England as *Unnatural Selection*. According to Weiss, the book's many ethnic and racial stereotypes may have been partially responsible for delaying its publication in the United States... "I tried to write it as a roach would write it," he explained. "These things leaped right out."

KAFKA ON COCKROACHES

"When Gregor Samsa woke up one morning from unsettling dreams, he found himself changed in his bed into a monstrous vermin." So begins Franz Kafka's *The Metamorphosis*, first published in Germany in November 1915. But was this monstrous vermin a cockroach, as so many readers have assumed?

The author of this existentialist classic was tight-lipped. Kafka forbade any illustration of Gregor's transformation from appearing on the book's cover, specifically informing his publisher that "the insect itself cannot be drawn." In accordance with this decree, illustrator Ottomar Starke created a cover for the book's first edition that depicts a man in a dressing gown and slippers, his hands shielding his face in fear. A short distance from this man is an opened door, through which one can only see the blackness of the abyss.

Kafka's careful choice of words to describe his insect were intentionally vague. Gregor is described in the original German text as an *ungeheueres Ungeziefer*, the first word connoting "a creature who has no place in the family," the second "an unclean animal not suited for sacrifice." In addition to "monstrous vermin," this alliterative phrase has been translated as an "enormous bug" and a "gigantic insect." Not many clues here.

> *"When Gregor Samsa woke up one morning from unsettling dreams, he found himself changed in his bed into a monstrous vermin."*
>
> —FRANZ KAFKA, The Metamorphosis

So the question remains: exactly what kind of an insect did Mr. Samsa become? In the earliest stages of his transformation, Gregor is said to be flat like a bedbug, so thin that he can easily slip beneath his bed, yet long enough to manipulate the door's lock with his teeth. The text tells of his "vaulted brown belly, sectioned by arch-shaped ribs," and "his many legs...waving helplessly before his eyes." Much later, the Samsa family's cleaning woman calls Gregor a *Mistkafer*, literally interpreted by most scholars as an "old dung beetle," another name for the scarab beetle. This has inspired several critics to draw connections between the symbolism of *The Metamorphosis* and the ancient Egyptian belief in the afterlife, which was represented by the scarab (whose larvae hatch below ground and emerge from the dry, lifeless desert soil).

We may never know if Kafka really had a roach in mind. Shortly before his death at age forty-one, he wrote to the novelist Max Brod, requesting that all of his papers, including the manuscripts to his novels, be burned. Brod chose to disregard the dying writer's wish; many years later, he edited a collection of Kafka's diaries. No mention was made of Gregor's true identity in any of these works. In all likelihood, the cryptic main character of *The Metamorphosis* represented none other than the estranged and alienated author, who, like many roaches, was tormented by humanity throughout much of his brief life.

Cockroaches on the stage

In the 1950s, Archy's texts were set to music and performed by an all-star cast, which included Carol Channing, John Carradine, and Eddie Bracken. Comedian Mel Brooks assisted Joe Darion in writing the script for this "back-alley opera," which opened off-Broadway as *archy and mehitabel*. Two decades later, this musical was reborn as the animated film *shinbone alley*, directed by John D. Wilson.

Surprisingly, this was not the first time that a cockroach had taken center stage. Among the lost works of the great Spanish poet, Federico Garcia Lorca is said to have been a mixture of narrative verse and dialogue telling the sad tale of a roach in love. According to Ian Gibson, the author of *Federico Garcia Lorca: A Life*, this insect is smitten by an injured butterfly that has fallen to the ground. Taken into the roach's home and tended by his family, the butterfly eventually recovers the use of its wings. It flies away, leaving the poor, brokenhearted roach to die.

Hearing this heart-wrenching recitation, Lorca's friends convinced him to transform the poem into a play, which the Eslava (a Madrid theater) would gladly produce. Lorca worked hard to finish the play, which, after last-minute hesitations about the title and the scenery, was finally staged as *El Malefico de la Mariposa* ("The Butterfly's Evil Spell") on an evening in March 1920.

Alas, *El Malefico* was a bomb. When the curtain went up, wrote Gibson, "neither La Argentinita, nor Catalina Barcena (in the role of the cockroach), nor Grieg, nor Mignoni's colourful set, nor Barrada's costumes, nor Martinez

Sierra's direction, nor the several merits of the little verse play itself, could overcome the rooted hostility of the audience."

The play's best lines were drowned out by a chorus of catcalls, insults, foot stamping, and witticisms. "A particularly ferocious outcry was provoked by the Scorpion's reference to his eating habits," Gibson continued. "When he exclaimed, 'I've just eaten a worm. It was delicious! Soft and sweet. Scrumptious!' Some wag shouted, 'Pour *Zotal* [a brand of insecticide] on him!' and the theatre rocked with mirth."

In the words of Lorca's biographer, "it was clear that Madrid was not yet ready (and no doubt never would be) for a verse play concerning the amorous misfortunes of cockroaches."

Roachy rhythms

As one might expect, most songs about roaches have been written by the people in closest association with them. A timeless classic is Albert King's version of *Cockroach*, originally released on the Stax label several decades ago. In this three-minute ballad, King complains about the complications of sleeping on a concrete porch. "Laying in my arms, where you ought to be, there's a big cockroach, looking up at me," he sings. Later, King moans that he's "always tired of these cockroaches crawlin' on up and down my arms and on my legs." The song ends with a heartfelt plea to "open the door" and let the poor blues singer in.

The Cockroach that Ate Cincinnati falls under the heading of a novelty song. Recorded by Rose and the Arrangement in 1974, it describes a fictitious B-movie monster, through the eyes of a self-professed horror film junkie:

> Frankenstein gives me the shivers,
> And Count Dracula's driving me batty.
> But they're not quite as bad,
> As the worst scare I've had:
> The Cockroach That Ate Cincinnati.

"Cincinnati" waxes poetic while portraying the monster cockroach's diet: "For lunch he'd just chew up a suburb or two; for dinner he ate the whole town (belch)." Clearly a crowd pleaser, it was included in at least one compilation CD, *Doctor Demento's Twentieth Anniversary Collection*.

My favorite cockroach song is by—who else?—the Roches, an all-female a cappella group from the Bay Area. As yet unreleased on an album, it appears in a Tiny Toon Adventure, performed by a trio of star-struck cockroach chanteuses, and includes such lines as:

> We are the Roaches, and if you think bugs aren't cool
> We knew the Beatles when they lived in Liverpool.

Also deserving mention are the fairly recent but largely overlooked *Roachmotel*, a title on the album *Intense Brutality* by Dead Youth; *Roaches*, by Bobby, Jimmy, and the Critters; and *Mi Brethren Roach*, by Jamaican reggae star Eek-a-Mouse. It should be noted that the "bugs" in Earl Hooker's instrumental *Two Bugs and a Roach* are tuberculosis germs, not insects. What the roach represents is anyone's guess.

Cockroach choreography

A catchy six-step, *The Roach (Dance)*, jumped onto the pop music charts in the early 1960s. It was performed by the dance's originators, Gene and Wendell.

"There's a dance, they call it the Roach; it's buggin' all the kids from coast to coast," the two-and-a-half-minute song opens, with Gene (or is it Wendell?) instructing everyone to "form a big line, to the porch," and get ready to "squish, squash, kill that roach." The roach, explains Wendell (or is it Gene?) has been buggin' him every day. However, one night while Wendell had to "beat it to the store to get some hair oil," he spied the roach walking with his girl. Both Gene and Wendell join in on the song's final chorus: "Let me at him (Squish squash!); I'm gonna kill him (Squish squash)."

In 1988, the Roach Dance was given new life by John Waters, the eccentric director of *Hairspray*, a feature-length film starring Divine (in a dual role) and introducing the then-portly Ricki Lake. The semiautobiographical *Hairspray* portrays the teen scene in Baltimore at the height of the Roach Dance's popularity. To make sure everyone used the right moves for the movie, Waters hired a well-known choreographer.

In the film's final moments, Lake appears in a silk evening gown embroidered with dinner plate-sized black velvet roaches. While co-stars Sonny Bono, Ruth Brown, Debbie Harry, and Jerry Stiller stare in stunned disbelief, she shows how to shake one's head, thorax, and abdomen to the sounds of Gene and Wendell.

"Stomp, step, skip 2, 3, 4, 5, 6, 7." The instructions to The Roach Dance are in the song.

"Squish, squash, kill that roach." Live roaches are not required, nor are rubber-soled boots.

Cockroaches as Pets and Prizes

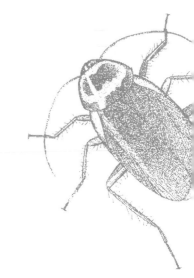

COCKROACHES CAN BE REWARDING animal companions. They're inexpensive to house, fairly resistant to disease, and far from finicky when it comes to food. They're not prone to howling at the moon or chirping, meowing, or squealing at dawn. Admittedly, they won't fetch your slippers. However, they won't need to be taken on twice-daily walks.

Like tropical fish, roaches are primarily pets to be contemplated, not played with. A few of the large, slow-moving species *can* be taken out and held, as one might handle a small pond turtle or a land hermit crab. Sure, some people have no desire to hold a three-inch-long crawling insect. However, many others can put aside their negative feelings and become better acquainted with one of the more fascinating invertebrates on our planet.

Cockroach tricks

You can't really train a pet roach. That's the word from Raymond A. Mendez, a cockroach consultant for such Hollywood films as *Creepshow* and *Joe's Apartment*.

"With insects, you're working with mobility machines, and what you're doing is more like a cattle wrangler who herds his cattle along," Mendez told a science writer for the *New York Times*. A better name for this line of work is "roach wrangling," he explained.

"Handling insects is like handling anything," says Mendez. "Pick up a cat the wrong way and it'll scratch you. But pick it up the right way, and it will snuggle in your arms and go to sleep."

The secret to holding roaches, he believes, is to know how much pressure to apply. Too little will allow the animal to escape, while too much will make it feel like it's going to be crushed, causing it to struggle in hope of breaking free.

It takes practice to acquire the right touch. "I tell people to handle their roaches in a bathtub, not in their living room, where they can escape," Mendez advises. "Play with your animals, letting them escape from you and gently recapturing them. It will take a while, but eventually you'll get the hang of it."

Having mastered this step, the vast majority of people (well, okay, pretty much everyone) still won't be able to teach their pets something they don't want to do. However, Mendez insists there are ways to steer insects so they appear to be performing on cue.

"For instance," he says, "roaches will normally run from light areas to dark areas. So if you construct a small wall with a dark hiding place on one end, then aim your cockroach at the wall and give it a slight tap to set it in motion, the cockroach will run along the wall toward the dark space, just like you—and the camera operator—had planned."

Using a similar strategy, entomology consultant Steven R. Kutcher created an unforgettable moment in the film *Race the Sun*—a long, uninterrupted shot of a roach climbing out of a tennis shoe, nimbly working its way across a bag of Cheetos, and stepping onto a surfing magazine, where it alights on a small color photo of a surfboard.

How did Kutcher pull it off? He first used a small, hand-manipulated plunger, hidden from view of the camera, to push the cockroach out of the shoe. Kutcher had previously made a crease in the bag of Cheetos, thus encouraging the scene's tiny star to follow its contours to reach the surfing magazine. To make the roach hit its mark on the magazine's color photograph, he used a blast of heat from a hand-held hair dryer. "Nothing to it," the roach wrangler modestly concludes.

"In May 1938, a prisoner in the Amarillo, Texas jail told how he had trained a cockroach to come to his solitary-confinement cell when he whistled. The cockroach would come with a cigarette tied to its back."

—Austin Frishman,
The Cockroach Combat Manual

Buying blattarians

As many as five different kinds of pet cockroaches can be purchased from the Carolina Biological Supply Company, one of the nation's largest retailers of life-science supplies. As of 1996, live Madagascan hissing cockroaches (a favorite of teachers and zoo docents) were selling for $7.50 each, three for $19.95. A complete habitat kit—a two-gallon plastic terrarium with an "escape-proof" lid, wood-chip bedding, food, water dish, climbing branch, set of instructions, and two Madagascan hissers—was priced at $28.95.

All-male and all-female batches of American cockroaches are available from Connecticut Valley Biological in Southampton, Massachusetts, priced around $28 a dozen. So are sexed and unsexed nymphs. Carolina Biological Supply and a few other companies also carry the ubiquitous German cockroach, plus some exotics like the death's head, giant, and West Indian leaf cockroach or drummer, *Blaberus discoidalis*.

Livestock swaps

The Blattodea Culture Group (BCG) hails from England. Its goal: "To promote the culture and study of worldwide cockroach species."

One of the BCG's primary functions is to facilitate the exchange of free "livestock," reared in captivity by its members. So if you've paid your membership dues and completed the necessary paperwork, you can pick and choose from over twenty cockroach species, many of which would make excellent pets.

The BCG's current membership includes collectors from Bermuda, Canada, France, Israel, Belgium, Finland, Malaysia, the United States, and Papua New Guinea. The group's secretary produces an irregular journal, with brief notes on cockroach care and feeding, and papers on aspects of blattarian biology. For additional information on the BCG, consult the Resources on page 173.

U-pick varieties

In many parts of the country, it's easy to find prospective pets without setting foot outside of your home. However, most of these insects will be the of the common "garden" variety—the Orientals, Americans, Germans, brown-bandeds, and smokeybrowns. To find more unusual species, one must start looking for roaches out-of-doors.

A cockroach safari is best mounted at night, during periods of peak blattid activity. While searching inside, all lights should be turned off to avoid making your subjects dash for cover. During hunts for peridomestic roaches, it may be more useful to leave any walkway, patio, or pool lights on, as these will attract certain species.

Choose a flashlight with a strong, focused beam. You can enhance its effectiveness by taping a round, yellow filter over its lens. A piece of #10 Roscolux theatrical lighting gel is ideal for this purpose. This will make the beam invisible to roaches, whose compound eyes are unable to detect the yellow band of the spectrum.

A long-handled mechanic's or dentist's mirror will also come in handy, helping you peer into cracks and behind barriers. Once they've located their quarry with a flashlight and one of these tools, professional exterminators use spray cans of repellent to flush roaches from such secret hiding places. However, this practice is not recommended for people who aren't trained and licensed to use these strong chemicals. A can of pressurized air, purchased at an art or photo supply shop, is a much safer bet. A well-directed blast from one of these cans is enough to make most of the Blattaria leave their lairs. Out in the open, these insects can be caught with a small dip net, transferred to a secure, well-ventilated container, and made into pets.

Adventures in the wild

A greater array of species can be collected from woodlands and fields. As with domestic species, these cockroaches are best observed and captured at night. A different set of tools may be needed to pluck these animals from their outdoor habitats. Particularly valuable is a good headlamp, to free up both hands for nabbing roaches on the run.

Where should one look for wild cockroaches? Collection sites for the various species listed in *Catalog and Atlas of the Cockroaches of North America North of Mexico* reveal a broad range of locality types:

> "This pestiferous race of beings, says an observer, are equally noisome and mischievous to natives or strangers, but particularly to collectors."
>
> —author unknown, *Natural History of Insects*, vol 2. (1830)

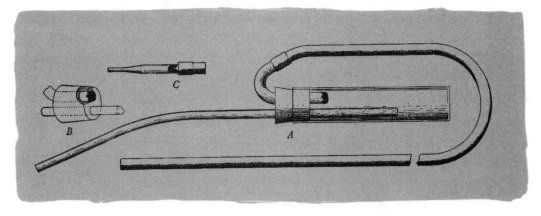

Design for a simple aspirator courtesy of the USDA.

"cool, deep ravines along the Appalachicola River"
"burrows of the kangaroo rat in southern California"
"humid, montane lowlands in the far western United States"
"scrub, flatwoods, and hammock communities in
        northeastern Florida"
"hollow twigs in mangroves"
"around residences in Houston, Texas"
"the extensively landscaped grounds of several motels"

"The small roach nymphs can be picked up with an aspirator, but the larger nymphs and adults must be caught with a pair of tweezers or the fingers," wrote Lafe R. Edmunds in 1953, describing his experiences with collecting and culturing native wood cockroaches in Ohio. Edmunds' aspirator was probably a gently curved glass pipe, with which the collector carefully sucked up his prey. A more sophisticated design for an aspirator can be assembled from two plastic straws, some rubber tubing, and a plastic or glass vial— to reduce the possibility of swallowing what's inhaled.

"Wood roaches will usually run rapidly when the cover under which they are resting is removed by the collector," Edmunds observed. But Ohio's cold winter and early spring weather seemed to slow overwintering nymphs, making them much easier to catch.

Traps

Roaches that are difficult to collect by hand are often easily caught with traps. Some roach traps rely on simple but time-tested materials. Others employ laboratory-synthesized sex pheromones or state-of-the-art baits to lure tiny game.

You can make your own roach trap by placing a slice of white bread into a quart Mason jar. Baby-food jars can also be used. Their small sizes and squat shapes make them easier to place in tight spaces. Coat the inside of the jar mouth with a slippery two-inch-wide band of petroleum jelly (the real pros mix three parts of this jelly with one part mineral oil). Now dig a small hole with a garden trowel, and sink the jar so that the top is flush with the surface of the ground. Cover the trap with a piece of wood, kept a half inch above the ground by pebbles or twigs. Roaches that walk under the wood and into the trap will stay there until someone takes them out.

In the log of the infamous H.M.S. *Bounty*, Captain William Bligh recounted his battles with blattarians, which threatened to devour the precious breadfruit plants aboard his vessel.

Rules for cockroach collectors

There are no formal bag limits on cockroaches. However, collectors should recognize the potential impacts of harvesting large numbers of specimens from the wild. The ability of a population to withstand such losses should be determined before any individuals are removed from their natural settings.

The sale and purchase of Madagascan hissing cockroaches is prohibited in Florida, a state already plagued by an assortment of accidentally introduced insect species. Importation and interstate transport of cockroaches is regulated by the United States Department of Agriculture's Animal and Plant Health Inspection Service. Before roaches can be shipped out of state, a permit must be obtained from this government office, located in Riverdale, Maryland.

Your pets' new home

An indoor cockroach enclosure must be comfortable for its occupants and secure and attractive for its owner. The best way to satisfy all of these criteria is to purchase a glass fish tank or terrarium, ten gallons or larger, and a snug-fitting screen top. As an extra safeguard against escape attempts, coat the top two inches of the tank with a commercial lubricant such as Stick 'n Slick® or a band of Teflon® insect-barrier tape.

Cover the floor of the tank with wood shavings, vermiculite, or an absorbent animal bedding like Alpha-Dri®. Now decorate your "roacharium" with tree bark (for Madagascan hissing, death's head, wood cockroaches, and other forest species), or toilet-paper tubes and egg cartons (for American, Australian, and other domestic species). In addition to making their habitats more visually appealing, these decorations provide your pets with places to hide or hang out during the day. They also encourage animals to stay put, making them less prone to attempt an escape.

Because most roaches come from tropical countries, they must be kept moist and warm—at least seventy-five degrees Fahrenheit to be happy. If you're not willing to give your pets their own heated room, purchase a heating pad from a pet store. Several different styles are available, primarily designed for snake, lizard, or turtle terrariums. The best ones for cockroaches fit under their tank and cover half or two-thirds of the bottom. This provides a gradient of heat, giving the tank's inhabitants some freedom of choice.

Captive cockroach cuisine

Offer your new pets a mixture of grains, vegetables, and fruit. These items don't need to be fresh, but sufficient quantities should be provided to prevent the roaches from nibbling on their enclosures or each other. Many cockroach culturists swear by kibbled dog and cat food, while others recommend Purina Cricket Chow®—a mealy ration made of corn, wheat millings, soybean hulls, fish meal, animal fat, molasses, and copious quantities of vitamins and minerals. I have maintained my pets for many months on diets of high-quality flake fish food, more easily nibbled than dried dog or cat food and more readily acquired than specially formulated cricket chow.

K. P. F. Haskins of the Pest Infestation Laboratory in Bucks, England, has published his recipe for making your own American cockroach feed: rolled oats, wheatfeed, fishmeal, and dried yeast powder in the ration of 9:9:1:1.

This food, according to Haskins, must be sterilized at 122 degrees Fahrenheit (50 degrees C) for fifteen hours before use.

All roaches are moisture dependent, so reliable sources of water are essential elements of any display. Unfortunately, providing your pets with a big bowl of water will, in all likelihood, create a serious drowning hazard. For this reason, it is better to make or buy a simple watering device. The least complicated of these is a two-inch square of sponge in a shallow dish or jar lid filled with water. Almost as low-tech are store-bought chick waterers, the bases of which can be filled with pebbles to keep your cockroaches from wading over their heads. A more advanced system involves drilling a hole through the lid of a small plastic petri dish, and inserting a roll of dental cotton through the hole, so that two inches project into the dish and one-half inch sticks out. When the dish is filled with water, the cotton acts like a wick, providing a constant supply of clean water. All three devices must be cleaned and refilled regularly.

Viewing aids

Typically, cockroaches like to hide by day. So if you're seeing much of your blattarians during these hours there is probably something wrong. Most insect zoos keep their blattarians in overly crowded tanks with not enough places to hide, thus ensuring that the animals will stay in the spotlight during prime viewing hours. In your home, however, they should be given ample niches and plenty of space. The surest signs of success in a tank are roach births. When the nymphs mature into adults, you can rest assured that you've done everything right.

While somewhat pricey, a good illuminated magnifier is indispensable for watching your pets at night. The crème de la crème of these instruments is an illuminated Coddington Magnifier, which projects its battery-powered beam through an optically corrected, 10-power Bausch & Lomb® lens. However, the price of one (around thirty dollars) can be an obstacle. A number of more affordable models with a plastic lens are available for less than ten dollars. Again, consult page 173 for information on ordering these.

Mass production

Millions of roaches are reared each year in laboratories across the United States. Not destined to be pets, these animals will become living crash-test dummies, subjected to hundreds of unusual and often cruel experiments conducted by researchers from both the private and public sectors.

Many of these experiments involve vivisection; mutilation and mass killing are not uncommon. Nonetheless, much of our understanding of modern neurobiology comes from such experiments. The thousands of insect deaths are also instrumental in our unflagging efforts at developing effective cockroach controls.

To ensure a steady supply of specimens for testing their line of products (which includes the time-tested pesticide Raid®), the S. C. Johnson Corporation breeds as many as eighty thousand roaches in a week. At its Wisconsin research facility (nicknamed the Raid Research Institute), full-time staff tend to the needs of several million nymphs and adults at a time. S. C. Johnson workers rear various age classes in eight-inch plastic rearing containers, each supplied with a screen top. Nymphs—as many as five thousand per container for some age classes—are fed Mazuri® rodent pellets and given water containing a drop of household bleach to keep it fresh.

In such tight quarters, cockroach cannibalism can be a problem. The United States Department of Agriculture's Medical and Veterinary Entomology Research Laboratory in Gainesville, Florida, estimated that it was losing up to 95 percent of each of its laboratory broods to these in-house attacks. However it solved this serious problem in 1994 by installing rearing cages with screen doors. The mesh in these doors is too fine for the adult females, some 400 to a cage, to pass through. It's just right for their newly hatched nymphs, which can escape to an adjacent cage. In their new home, the nymphs don't bother each other, because they are all the same size and age. They are free to molt and grow in peace—at least until the day they are dispatched in the name of science by the Department of Agriculture.

> Among Thomas A. Edison's many patented inventions was a device for electrocuting cockroaches

Cockroaches in zoos and museums

For some reason, the London Zoo and several other large institutions believe that people will line up just to see a really spectacular roach-infested kitchen. They're right. The London Zoo's ultra-realistic American cockroach exhibit is one of the most popular attractions at this historic facility, one of the oldest zoological gardens in the world. It features a cutaway sink with dirty pots, pans, and plates, and some strategically placed props—a bottle of dish detergent, a carton of milk, and a half-eaten apple—to impart authenticity.

"On special occasions such as Christmas, the display is decorated accordingly," explained insect keeper Paul Atkin. "The cage is dimly lit and looks very 'atmospheric' with a heaving mass of bodies and waving antennae. Add to this the odd one or two, occasionally seen running around the sink. We like to think the display has never looked the same way twice in its life."

A simulated kitchen crawling with real German cockroaches is also a central attraction of The Insectarium, founded by Steve and Karen Kanya, owners of Steve's Bug Off Exterminating Company in Philadelphia. The Kanyas say they came up with the idea for their insect museum by watching people stop at Bug Off's window to admire the "catch of the day"—an impressive rat or some unusually large roaches contained in a ten-gallon tank. "People would walk by and stop dead in their tracks," Mr. Kanya said. "They always wanted to see more."

More is what they got. The Kanyas invested $150,000 to renovate the top two floors of their business, and The Insectarium opened its doors in January of 1991. Now more than fifty thousand people pass through this novel facility each year, in order to get glimpses of not only cockroaches, but giant waterbugs, tarantulas, Goliath beetles, and other champions of the arthropod empire too.

Roaches have also been freed from the confines of our kitchens and placed in lush jungle settings at a number of museums and zoos woldwide. A few species, including the death's head cockroach, are regular crowd-pleasers featured in the popular—and timely—tropical rain forest exhibits recently added to several zoos. Randy Morgan, curator of the Cincinnati Zoo Insectarium is especially proud of his strikingly patterned Trinidad zebra cockroaches (*Eurycotis decipiens*), and at Seattle's Woodland Park Zoo, Madagascan hissers are always a hit. While visiting the Ralph Parsons Insect Zoo at the Natural History Museum of Los Angeles County, look for American cockroaches—not the ones on display, but the "wild" kind, snug as the proverbial bug in the museum's restrooms.

Butterfly gardens at the San Diego Wildlife Park, the Houston Museum of Natural Science, and several other locations across the United States and Canada are helping to win some respect for all manner of insects. It's possible that live roaches will be even more prominently displayed by museums, zoos, and wild animal parks in the future. Information on existing cockroach displays here and around the world can be found on page 172.

Cockroach collections

Not to be mistaken for cockroach preserves, preserved cockroaches are the real trophies on many an entomologist's specimen shelf. And it doesn't take much to acquire some of these gems—just a bug-killing jar, some liquid preservative, and a bottle. If you'd like to create a more impressive display, some steel pins and a foam mounting board are all that's required.

JAR

To make your own killing jar, pour half an inch of plaster of paris into the bottom of a wide-mouthed jar. Allow it to dry overnight. Then wrap the bottom third of the jar in electrical or duct tape, to reduce the chance of your jar shattering when dropped. Immediately before using your jar, sprinkle some of the ethyl acetate or nail polish remover onto the plaster of paris. Quickly close the jar's lid. The roach-dispatching vapors will remain effective for a few days.

VAPORS

No material that can rapidly asphyxiate insects could possibly be all that healthful for humans. Even the safest—either ethyl acetate or fingernail polish remover—should be cautiously approached. Avoid breathing their fumes or dousing your clothes or skin with them. To prevent any tragic mistakes, label your killing jar with a single word: POISON.

ACCESSORIES

What's to be done with a roach after it's been captured and killed? It's simplest to put any dead specimens into a tightly capped bottle of preservative, and label the bottle with any pertinent data—genus, species, date collected, where caught, and so on. Mounting specimens is extra trouble, but it yields considerably more interesting results. A single pin may be all it takes, piercing the right tegmen about midway through the body. Other pins can be strategically placed to hold open the cockroach's wings while they dry in a fixed position. For unusually large or robust specimens, cardboard supports or additional pins may be needed. A detailed set of instructions for mounting insects is presented in *The Practical Entomologist*, by Rick Imes.

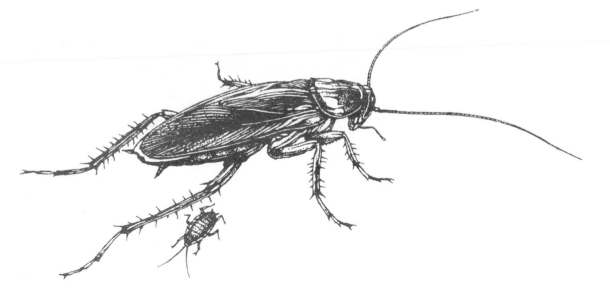

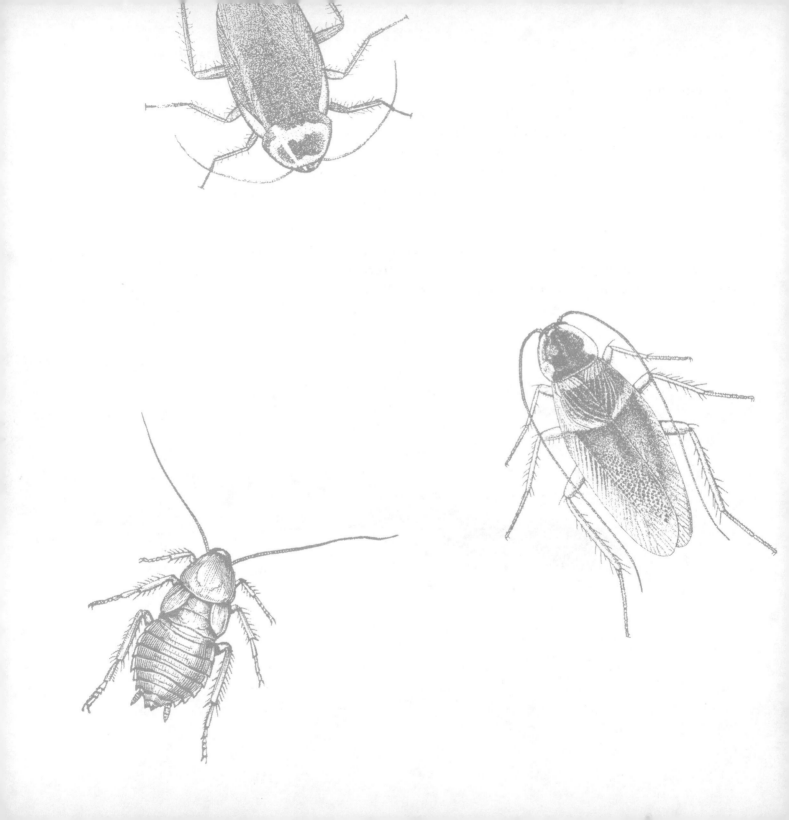

Can We Control Cockroaches?

THE EARLIEST REMEDY FOR COCKROACH infestations is contained in the *Book of the Dead*, a seventy-two-foot-long papyrus scroll dating back to the eighteenth dynasty (1750-1304 B.C.) of Egypt. Arranged in vertical columns within this document are hieroglyphs describing at least 200 spells. The majority of these spells deal with assisting the spirits of the deceased in the afterlife. However, many are closely focused on the here and now. One of these is said to have been first cast by the ram-headed god Khnum—the shaper of the sun, humankind, and the other Egyptian gods—perhaps to clean his celestial house at Elephantine, the mythical source of the Nile.

"Be far from me, o vile cockroach, for I am the God Khnum," this spell proclaims. This utterance may have also been used by officiating priests, who were required to fumigate areas with incense smoke, killing all insects before any of their work could be performed.

Drawings of roaches are altogether absent from the *Book of the Dead*. However, numerous images of scarab beetles—the Egyptian symbol for rebirth—appear in this work. It may be that high priests and sorcerers were afraid that the hieroglyphic symbols in their spells might come to life and scurry away, spoiling the magic. Then again, maybe the problem was that the roaches simply wouldn't sit still long enough to have their portraits painted. While roaches may have many virtues, patience isn't always one of them.

Archaic advice

Another written recommendation for removing roaches is attributed to the Greek scholar Diophanes, as published in Thomas Mouffet's 1658 masterwork, *The Theater of Insects*:

> Get the Guts of a Ram fresh killed and full of dung, bury it in the earth where many Moths [cockroaches] use, and cast the ground lightly upon it; two daies after all the Blats will gather to it; the which at your pleasure you may carry other where, or bury them deep enough in the place, that they shall not be able to rise again.

Mouffet, the Englishman, suggested that those afflicted with roaches "cast but a handful of Flea-bane...and all the Blats [cockroaches] will gather to it." In addition to fleabane (also called the "Blat-herb"), Mouffet's peers also believed that the silvery flocked leaves of *Verbascum blattaria*, a kind of mullein, would serve as a powerful repellent.

A simple but effective remedy appeared in the Memorial (1887) Edition of *Dr. Chase's Third, Last, And Complete Receipt Book and Household Physician*:

> I give a recipe to your correspondent who wishes to know how to get rid of the insects he calls the cockroaches, although I think he misnames them. Let his wife finish making peach preserves late at night in a smooth, bright, brass kettle; then persuade her it is too late to clean the kettle till morning, but set it against the wall where the insects are thickest and retire to rest. In the morning he will find the sides of the kettle bright as a new dollar, but he will find every insect that was hungry in the bottom of the kettle, when, if he uses the recipe I did, he will treat them to a sufficient quantity of boiling water to render them perfectly harmless. As I thought molasses cheaper than peach preserve juice, I ever afterward baited the same trap with molasses, and I caught the last one of millions.

Even less trouble is Vernon Haber's technique, which appeared in *Cockroach Pests of Minnesota with Special Reference to the German Cockroach*, a 1919 publication of the University of Minnesota's Agricultural Experiment Station:

> At night, put old cloths dampened with dish water in the sink or near their runways and places of seclusion. Darken the room and leave it. At half- or three-quarter hour intervals, return with a liberal supply of scalding hot water and dash it upon the cloths, thus destroying many cockroaches which have secluded themselves in the folds of the cloth or beneath it. The dead cockroaches should be collected and burned before the cloths are rearranged to trap more.

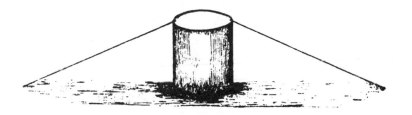

In *The Living American House*, George Ordish describes a patented device, Johnson's Perfect Roach Trap, from the mid 1800s—"a balanced metal leaf gave way from the weight of the cockroach, and precipitated into a vessel from which it could not escape." A handful of alternative designs are presented in Herrick's *Insects Injurious to the Household and Annoying to Man*. One of these is simply a circular tin box with inclined runways. Another, more complicated but "most serviceable," is comprised of "a small wooden box, having a circular hole at the top fitted with a glass rim, out of which it is impossible for them to escape." On the subject of bait for these traps, Herrick has written:

> Roaches in general are very fond of stale beer and advantage is taken of this, especially in England, to trap them by drowning them in this liquid. Any deep jar will serve for the purpose. It is partially filled with the beer and sticks are then inclined against the jar on the outside and bent over until they project into the beer. The roaches climb up the sticks and slip down into the liquid in which they are drowned.

Ridiculous remedies

Innovative but ineffectual methods for cockroach control abounded in the eighteenth, nineteenth, and early twentieth centuries. For example, Frank Cowan's *Curious Facts in the History of Insects*, an idiosyncratic compendium of factual and fanciful beliefs from 1865, contains the following three outlandish suggestions:

> A quite common superstitious practice in order to rid a house of Cockroaches, is in vogue in our county at the present time. It is no other than to address these pests a written letter containing the following words, or to this effect: "Oh, Roaches, you have troubled me long enough, go now and trouble my neighbors." This letter must be put where they most swarm, after sealing and going through with the other, customary forms of letter writing. It is well, too, to write legibly and punctuate according to rule...

> Another receipt for driving away Cockroaches is as follows: Close in an envelope several of these insects, and drop it in the street unseen, and the remaining Roaches will all go to the finder of the parcel...

> It is also said that if a looking-glass be held before Roaches, they will be so frightened as to leave the premises.

Even stranger pieces of advice were somehow overlooked by Cowan. One from nineteenth-century Mexico tells the roach-afflicted to "catch three cockroaches and put them in a bottle, and go carry them where two roads cross. Here hold the bottle upside down, and as they fall out repeat aloud three *credos*. Then all the cockroaches in the house whence these three came will go away."

Another, presented in Lucy Clausen's *Insect Fact and Folkore*, is a variation of the "sealed envelope" idea. Said to have hailed from Springfield, Massachusetts, it advises those at wit's end to "catch a roach, put it in a piece of paper with a small amount of money, give the parcel to anyone who will take it, and the cockroaches will go to the house of the person who accepts the parcel."

Within a week of publishing a cockroach-control recipe in 1982, the editors of *Hints from Heloise* received 40,000 stamped, self-addressed envelopes from readers requesting copies of this advice.

Shipboard solutions

Remedies for shipboard infestations, while ambitious, were often to no avail. A standard treatment for commercial vessels in Victorian England involved evacuating all passengers and crew, placing pots of lighted sulphur in the hold, and battening down the hatches for twenty-four hours.

Aboard other vessels, hand-picking was the preferred alternative. Prior to World War II, seamen in the Japanese Navy were granted a day of shore leave for every 300 cockroaches they captured. The *Danish Navy Annals* of 1611 describes shipboard hunts, with a bottle of brandy from the cook's locked pantry as the reward for every thousand blattarians caught. The *Annals* also tells of a single shipboard search, during which some 32,500 roaches were bagged. No mention is made of any difficulties keeping the Danish ship on course, after all thirty-two bottles of brandy were doled out!

Hedgehogs for hire

A thorough description of the hedgehog's role is contained in *Bingley's History of Animated Nature*, a two-thousand-page treatise from 1868:

> A gentleman, whose kitchen in London was infested with black beetles [a.k.a. Oriental cockroaches], was recommended to put a Hedgehog in it. He, consequently, had one brought there which had been caught in his garden in the country...In a little time, it became so domesticated as not to fear either cats or dogs; and even to take its food out of the hand of any one who offered it...By his good services he well merited his board and lodging, for scarcely one beetle was left in the house; and it is supposed that he also destroyed the mice.

According to Bingley, this meritorious hedgehog was kept in an upright basket, which at the family's bedtime was placed in the infested kitchen. Its insectivorous diet was supplemented nightly with soaked bread and a pan of milk. Such nutritious fare caused the hedgehog to grow so fat that, according to Bingley, "after a little while, it was with difficulty he could squeeze himself under the closet-door."

Several breeds of hedgehogs are currently available from pet shops and exotic-animal suppliers. Six- to eight-inch-long African pygmy hedgehogs are the most readily available of these. Like Bingley's European hedgehog, this species is basically nocturnal, reaching a peak of activity between nine in the evening and midnight. This puts the African pygmy in an excellent position to meet and eat roaches. It should be noted, however, that these foreign hedgehogs have the same potential for mayhem as any exotic pet: if allowed to escape, these introduced species can settle and (under favorable conditions) proliferate, forcing out their native counterparts. For this reason, hedgehog owners must accept full responsibility for their pets, making sure that such accidental introductions will never occur.

Altering the environment

They don't call them the exterminator's bread and butter for nothing, as it may take repeated visits by the "hired guns" to rid one's home of these insects. But there are many things that you can do in your home to discourage roach populations from building. Among these, my least favorite measure—fastidious housekeeping—has proven to be the most effective anticockroach assurance.

FOOD

The way to a roach's heart is through its stomach. So get rid of all accessible sources of food. Grains and other dried staples should be stored in plastic or glass containers with tightly fitting lids. Overly inviting fruit bowls and candy dishes should be covered or eliminated altogether. Dog and cat food should be served within a second, shallow container of soapy water—an unfordable moat for roaches.

WATER

Eliminating sources of water available to them will greatly speed the demise of cockroaches that you are trying to starve out. After all obvious plumbing leaks have been fixed, search for smaller cockroach oases that can be eliminated—sink and bathtub drains without stoppers, opened but not entirely empty beverage containers, or drain pans under refrigerators and air conditioners. Don't overlook dish pans and fish tanks. For every drop of drinking water taken out of circulation, you may be ridding yourself of at least one unwanted houseguest.

GARBAGE

Trash piles become all-night diners for cockroaches. At the very least, remember to put a lid on any garbage can that might provide nourishment; better yet: keep all of your food waste in one can, depositing a two-inch layer of sawdust over each new layer of food. The whole thing can then be taken outdoors, mixed with grass clippings, straw, or leaves, and allowed to decompose, producing a fine grade of compost for the garden.

HARBORAGES

Get rid of roach hangouts. Small, nymph-sized crevices can be sealed with a thick coat of paint; larger ones may require caulk or putty. Steel wool crammed into cracks around ductwork or water pipes will compensate for the "open sesame" effect—the expansion and contraction of these openings as they warm up and cool down. Before sealing off harborages, vacuum and wash them down to remove any egg capsules or feces.

HEAT

Freeze 'em out. Just by lowering the temperature of your apartment or home, you can slow growth rates and extend the gestation periods of pest roaches. You'll be battling even fewer roaches if you turn the heat off entirely. Charles L. Marlatt described one freak winter in Florida in 1894, which "destroyed all the cockroaches, even in houses, except a few unusually well protected." Not many roaches can survive prolonged exposure to temperatures below freezing.

OUTDOORS

The same rules for indoor populations apply to the control of peridomestic cockroaches outside. It is just as important to clamp down on sources of water and food and to eliminate as many harborages as humanly possible. This may entail removing layers of wood chips or other bedding materials from outdoor plantings and adding screening to any vents in a home's walls or foundation. To control the extremely prolific Pacific beetle cockroach, the Hawaii Cooperative Extension Service recommends spraying the trunks of shrubbery with the over-the-counter insecticide diazinon, applied at the rate of one teaspoon of 25 percent emulsifiable concentrate per gallon of water.

Try a few traps

Many kinds of sticky traps are sold at hardware and grocery stores. Most are made of cardboard, shaped into rectangular boxes or triangular tubes, open at both ends. Their interior surfaces are coated with sticky glue, and some brands contain a small strip of scented bait. "The roaches can check in, but they can't check out," proclaim the ads for Roach Motel®, one of the more popular ready-made traps.

Because state-of-the-art, baited traps are relatively expensive, the budget-conscious may prefer brands without bait. A few drops of banana extract applied to the inside of one of these budget traps will greatly enhance its blattarian appeal. Even cheaper to acquire and set, mason and baby-food jar traps work equally well indoors; they can be modified for indoor placement by adding an exterior coat of black paint or a paper wrapping to darken their interiors.

Mason jar traps can be nearly as effective as store-bought sticky traps. In 1980, Virginia Polytechnic Institute entomologist Mary H. Ross compared the catch potentials of mass-produced "roatel" traps and homemade jar traps. She found that both caught roughly the same numbers of adults and older nymphs. However, roatels were far superior at snagging small nymphs, which may not be capable of scaling Mt. Mason Jar's slippery outer surfaces. Over a two-week trial period, one roatel snagged a total of 364 German cockroaches—103 more than the jar trap.

"As evidence of its [the German Cockroach's] abundance under favorable conditions, it may be stated that in 1890 a single person captured for me over 30 adult specimens and fully half that number of young in less than ten minutes, in the kitchen of the leading hotel in the city of Terre Haute."

—W.S. Blatchley, *Orthoptera of Northeastern America, With Special Reference to the Faunas of Indiana and Florida* (1920)

A good trap will usually catch a few roaches in assorted developmental stages over a twenty-four-hour period. However, the right trap in the right place can snag fifty or more of these inquisitive insects in a night. A "no vacancy" sign on a Roach Motel® is not a pretty sight: the buildup of carcasses over a few days can actually prevent newcomers from entering a trap and sticking around, thus reducing the trap's overall effectiveness. Such a rich cache is no reason to get cocky: for every roach captured, there are probably several dozen eager replacements waiting in the wings.

No over-the-counter trap works equally well on all of the "big five" domestic pest cockroaches—the German, brownbanded, Oriental, American, and smokeybrown. The effectiveness of four brands was evaluated by W. S. Moore and T. A. Granovsky in 1983:

	G	B	O	A	S	NOTES
D-Con®	1	2	1	3	3	Molasses bait
Holiday Roach Coach®	2	1	1	2	2	Bait packet placed in trap by user
Mr. Sticky®	2	3	2	3	4	No bait
Raid Roach Traps®	1	2	1	1	1	Bait attractant
G=German, B=brownbanded, O=Oriental, A=American, S=Smokeybrown 1=best, 4=worst						

Proper placement of sticky or jar traps will ensure optimum catches. Cockroaches typically congregate in warm, sometimes wet niches, so traps should be positioned within five feet of such sites. Outside of their harborages, they tend to avoid open areas, preferring to peruse the perimeters of most indoor spaces. To nab these insects on the move, place several traps along interior walls, a few feet from a room's corners, so the cockroaches will be sure to stumble across them. Remember to put a few traps at various elevations, as brownbanded cockroaches and a few other heat-loving species are often found on shelving, behind framed pictures, and in the upper folds of draperies.

No sticky traps, regardless of their "species appeal" or placement, will eliminate all cockroaches from an indoor environment. However, by regularly counting your catches, you can use data from trapping to identify population trends and patterns of dispersal—and to determine if you are winning or losing your war against these indoor pests.

COCKROACH SPECIES	G	B	O	A	S
Average number/day/trap: low infestation	5	3	1	1	1
moderate infestation	5-20	3-10	1-10	1-10	1-10
high infestation	20-100	10-50	10-25	10-25	10-25

G=German, B=brownbanded, O=Oriental, A=American, S=Smokeybrown
Adapted from Cockroach Control Manual, Lancaster County (Nebraska) Cooperative Extension

Guardian geckos

With a cult following in the Caribbean and near-totemic status in Hawaii (where it's considered taboo to harm one), the gecko has quickly become the homeowner's roach-catcher of choice. Around seven inches long including the tail, and rather attractive (its sand-colored body is decorated with wavy chocolate-brown bands), the African house gecko is both solitary and secretive. It hunts at night and holes up by day, usually on the upper shelves of a cabinet, behind a refrigerator, or in some other warm, safe, and out-of-the-way spot.

Geckos can walk up walls and across ceilings with ease. This feat is made possible by their toe pads, which are equipped not with suction cups but with rows of microscopic, velcrolike hooks. These hooks can latch onto any surface irregularity, regardless of its size, even on glass.

"Like tiny tigers stalking prey, they creep along," wrote Kenneth Petren and Ted J. Case in *Natural History* magazine, " keeping their bodies pressed close to whatever they're walking on, and finish off the hunt with a final pounce and snap of the jaws." Reports from owners of silent-but-deadly geckos are truly impressive. The author has spoken with several people who were actually forced to release crickets in their homes—as gecko food, after the populations of roaches had been polished off.

As exotherms (cold-blooded animals), all 650 species of geckos draw heat from the sun. Larger species are able to take in more heat and hold it longer, making them better suited for house and apartment life in temperate zones. This is the main selling point for the tokay gecko, a substantially larger animal than the house gecko. Unfortunately, this bright reddish purple species is also considerably more obtrusive, with a loud, throaty bark (on a par with a small poodle) and a deservedly bad reputation as a finger biter.

Because neither gecko species is native to North America, the consequences of an accidental introduction must again be considered before any specimens are purchased to be released indoors. This is particularly true in Florida, Texas, and the other southern coastal states, where conditions are most favorable for escapees. In Hawaii, nonnative house geckos are starting to outcompete less-robust fox geckos and mourning geckos—as they have previously done in Fiji.

Abhorrent herbs

The recent resurgence of interest in "green" alternatives to chemical cures has inspired renewed interest in herbal remedies. Liquid extracts of lemongrass, peppermint, basil, lavender, citronella, angelica, and a few other plants are said to inhibit cockroach foraging patterns. Sold by health-food stores and mail-order businesses as "essential oils," these herbal products can be dabbed on cotton balls and stuffed in nooks and crannies. *The Insect Free Herbal Household* recommends diluting four drops of oil in a pint of warm water, then spraying the mix on the cockroaches' favorite crawl spaces.

Word of mouth is generally supportive of such herbal treatments, however the scientific record is surprisingly quiet on this subject. Some researchers believe essential oils may at best serve as temporary roadblocks, requiring cockroaches to seek new avenues of egress. As the pest species eventually become accustomed to a room's strange, new scents, the users of botanical blatticides may be required to adopt more traditional chemical controls.

Their effectiveness aside, many botanicals are costly. For example, essential oil of citronella sells for fifteen dollars a fluid ounce; essential oil of sweet basil sells for nearly twice as much; both oils may be prohibitively expensive when treating large surface areas. However, their pleasing aromas and negligible effects on water quality or soils make them a welcome addition to the anticockroach arsenal.

Presently, the trendiest herbal repellent is the yellow-green fruit of the Osage orange tree. As big as a grapefruit, one of these thick-skinned orbs is said to emit a strange perfume that repels roaches. A number of people claim to have gained the upper hand by placing the fruits under their beds, in closets, and in cabinets beneath kitchen sinks.

New Yorkers have staked out some of their region's larger Osage orange trees. A magnificent fifty-foot specimen is located near Driprock Arch, just west of the Wollman skating rink in Central Park. "I went back to visit it last week in search of some free antiroach balls, but the ground had already been scoured and the remaining few were too high to harvest," a woman complained in the *New York Times*.

FLOWER POWER

Plants produce toxins for one reason—to discourage animals from eating them. People have been gathering these poisons for centuries, concentrating their repellent ingredients in liquid and powdered forms to create natural insecticides. One of the most effective against roaches is pyrethrin, a derivative of the small, daisylike flowers of the bushlike pyrethrum plant. This commercially valuable plant is cultivated in east and central Africa, Brazil, Ecuador, and many other sunny locales.

In central California, the Buhach (pronounced "byew-hack") Plantation began raising pyrethrum plants as long ago as 1871. Pyrethrin-rich Buhach powder was originally sold to prospectors for fighting off ravenous Klondike mosquitoes during the Alaska gold rush of 1876. However, even after the short-lived epidemic of "gold fever," demand for the new multipurpose insecticide remained strong, attracting the interest of many different sectors over subsequent decades. In 1891, Clarence M. Weed's *Insects And Insecticides* printed the instructions for Buhach's use:

Just before nightfall go into the infested rooms and puff it into all crevices, under base-boards, into drawers and cracks of old furniture—in fact wherever there is a crack—and in the morning the floor will be covered with dead and dying or demoralized and paralyzed roaches, which may easily be swept up or otherwise collected and burned. With cleanliness and persistency in these methods the pest may be substantially driven out of a house, and should never be allowed to get full possession by immigrants from without.

While relatively harmless to vertebrates (especially when applied in low concentrations), pyrethrin powders aren't without side effects, as exposure to even low levels can trigger allergic reactions in some people. Nonetheless, pyrethrins have become more popular in recent years, perhaps because of the resistance of many pest species to synthetic insecticides. Reborn in 1981, the Buhach Company continues to do a brisk business, supplying the original formula of its "California Reliable Insecticide"—100 percent powdered pyrethrum flowers—in a bright yellow can.

Poisoned baits

It's a ploy that predates the Borgias: slipping a deadly poison in the favorite food or drink of an opponent. It still works nicely on cockroaches, providing the bitter taste of the death-dealing substances can be masked.

Unlike rodents and other so-called "higher" forms of life, roaches sample everything with their mouth parts before they ingest it. Any morsel that tastes the least bit tainted with chemicals will be rejected. Because of this, poisons must be subtly flavored and administered in small doses. No less important than taste is the way that a poison is served. To be effective, poisons must also be tantalizing—so tasty that roaches will choose them over the many dozens of other delicacies in their environment.

The best roach-killers combine a subtle, slow-working poison with some sort of food item that roaches will find alluring. Early formulas combined toxins from natural sources (such as wormwood and hemlock) with such tasty attractants as bacon grease, molasses, and bran. Shaped into balls or cakes,

161

these gob-stoppers would be placed in heavily infested parts of a house. Later baits contained white arsenic and phosphate paste, ingredients that not only killed cockroaches but any small children or household pets that happened to swallow them.

Today, relatively safe and tasty poisoned baits are available in such recent formulations as Combat® or MaxForce®. The active ingredient in both of these products is hydramethylnon, an inconspicuous stomach poison that, in extremely low doses, slowly finishes off insect diners two to four days after it is eaten. The bait is primarily corn syrup, mixed with various sugars to make it even sweeter. Sold in child-proof and pet-safe plastic bait stations, these products are among the more effective, environmentally friendly alternatives in the war against pest cockroaches.

RESISTANT ROACHES

Taste became an issue in 1983, when after several months of successful poisonings, Combat® cockroach bait started losing its clout. It wasn't that the cockroaches were becoming wise to the presence of poison. On the contrary, they had evolved a dislike for the taste of the bait.

Presumably, a few cockroaches always had a disdain for glucose, one of the main ingredients that Combat uses to lure its quarry. They steered clear of Combat® bait stations, surviving to sire offspring with the same glucose aversion. Cockroaches without this aversion ate the stuff and died without having any offspring. Eventually an entire population of glucose-hating cockroaches was turning its back on Combat®. Only by substituting pure fructose for the glucose could researchers bring Combat®'s kill ratio up. It's likely that Combat®'s formula will change again and again, in response to the cockroaches' fickle tastes.

In the same way that glucose aversion can be genetically passed from one generation to the next, cockroaches can also inherit resistance to insecticides of various forms. All it takes for this to happen is a handful of cockroaches developing a resistance to a certain formula. From this little group can emerge an entire strain of resistant cockroaches, invulnerable to one or more ordinarily lethal chemicals.

DUST TO DUST

Employed for more than a century, boric acid and its salts are still the most cost-effective roach killers for long-term use. And they are widely believed to be the safest for people and pets. So why are many pest control professionals so reticent to use them?

"The most frequent explanation for this reluctance is the fact that boric acid is generally applied as a dust, and most PCOs [pest control operators] have little experience with dusts and dust application equipment," is the explanation proposed by the authors of *Managing Cockroaches—The Least Toxic Way.* "They find such applications messy and time-consuming, compared to the more familiar liquid and aerosol formulations of conventional insecticides."

Pest controllers may also shy away from boric acid because of its slow killing action. Cockroaches must either ingest this fine white powder while preening themselves or absorb it through small pores in their cuticles. It can take one to two weeks before enough poison will have accumulated in the gut, causing the insect to die. This won't satisfy customers who want results right away.

Boric acid comes in a variety of forms, including powders, tablets, water-soluble washes, and aerosols. You should apply the powdered forms to the relevant surfaces as a thin dusty covering, as if you were lightly salting food. Rubber-bulbed dusters or plastic ketchup squirters are suitable tools for puffing the talcumlike substance into cracks and crevices. If you don't have either piece of gear, put a small amount of boric acid powder into a sealed envelope, tear a small hole in one corner, then gently squeeze its sides to force out a small amount.

Two other powders—silica gel and diatomaceous earth—are also commonly applied around homes. Unlike boric acid, these do not need to be ingested to work. Rather, their rough-edged particles become embedded in a cockroach's waxy cuticle, producing small scratches in its ordinarily impervious surfaces. The scratches become avenues for infection and water loss, causing the abraded blattarian to succumb to disease or desiccation in a matter of days.

Still stronger stuff

It's estimated that nearly $500 million is spent each year on insecticides to kill cockroaches in the United States, suggesting that Americans are as trigger-happy with aerosol cans as with handguns. Such a strong chemical dependency is nurtured by ambitious but unconscionable ad campaigns, which often portray cockroach poisoning as economical, safe, and downright fun. Only by reading the labels of most over-the-counter insecticide sprays is one likely to encounter the truth:

> Harmful if swallowed, inhaled, or absorbed through the skin. Avoid breathing vapors or spray mist. Avoid contact with eyes, skin or clothing. Do not allow spray to contact food, foodstuffs, and water supplies. Thoroughly wash food contact surfaces with soap and water if they become contaminated by application of this product. Do not allow children or pets to contact treated surfaces until spray has dried. Remove pets and cover fish bowls (tanks) before spraying. Do not use this product in edible product areas of food processing plants, restaurants, or other areas where food is commercially prepared. If swallowed, do not induce vomiting. Get immediate attention. If inhaled, remove from contaminated atmosphere. Give artificial respiration and oxygen if necessary. This product is toxic to fish, birds, and other wildlife. Do not apply directly to water. Do not contaminate water by cleaning of equipment or disposal of wastes.

Because even our pest species are elements of an intricate ecological system, any efforts to completely eliminate them will have an effect on many other forms of life. Among the innocent victims of wholesale insecticidal applications are the many beneficial insects, spiders, and centipedes that grace our homes and gardens. Without these creatures to keep mites, aphids, and other serious agricultural and domestic pests in check, stronger chemical controls are often required.

Many insecticides create more serious problems outside of their immediate sites of application. When birds and other animals eat the tainted carcasses of insects killed by these compounds, they, too, can fall victim to lethal

In the parlance of licensed pesticide appliers, the moment that a poisoned cockroach starts twitching is known as its "knock-down" point; an "irreversible knock-down" is when a poisoned cockroach remains on its back for two or three minutes without struggling, then dies.

chemical effects. Certain insecticides are remarkably persistent in the environment; outlawed in the 1960s because of its effects on wildlife, DDT can still be found in its residual forms in soils and water supplies throughout the United States. The combined effects of less blatantly harmful insecticides (with such sinister Dr. Strangelove-esque names as Fenitrothion, Bendiocarb, and d-Tetramethrin) have yet to be fully understood.

A puzzling death pose

When a cockroach drops dead, why does it always land on its back? This question, originally posed to Cecil Adams, the creator of the tell-all newspaper column, *Straight Dope*, merits a long and thoughtful reply. One possibility for the belly-up pose, according to Adams, is that the cockroach has suffered a heart attack while climbing a wall.

"Just suppose the roach expires somehow and tumbles earthward," he wrote in one of his columns. The aerodynamics of the cockroach body—"smooth on the back, or wing side; irregular on the front, or leg side"—are such that the dying insect would tend to land on its back.

On the other hand, the cockroach in question may have passed on after ingesting potent neurotoxins. These neurotoxins, Adams explains, cause the roach to "twitch itself to death," inadvertently flipping itself over onto its back, where it can only "flail helplessly until the end comes."

In a third scenario, the cockroach may have desiccated and come to rest on its belly. No more than a hollow shell, it would easily be blown onto its back by a passing breeze.

All three possibilities seem plausible to entomologist Michael Rust, who hastens to add that cockroaches can't have heart attacks. Rust also cautions that cockroaches *don't* always die on their backs. When death is slow, as it often is with boric acid or poisoned baits, a stricken insect may crawl back to its harborage and assume one last lifelike, legs-down pose. Of course, people are less likely to find the corpse of a cockroach that has passed away in this fashion.

In 1934, Spain issued a set of stamps to commemorate the 300th anniversary of the death of Lope de Vega. Two of the stamps bear images of the great author; a third is a still life, said to have been taken from de Vega's bookplate. It pictures a dead cockroach, flat on its back—symbolizing one of de Vega's critic, so the story goes.

Novel devices

At least a dozen innovative technologies for cockroach destroyers have been developed in the last fifteen years. The most recently unveiled are new strains of the microbe *Bacillus thuringensis* (or Bt), genetically engineered versions of already familiar garden-pest destroyers. This and every other new addition to the blatticidal arsenal have been greeted with enthusiasm by the press. Several have become welcome additions to the stockpile, while others have gone straight to the scrap heap after a few months of field tests.

KILLER FUNGI

Perfected in 1993, the Bio-Path® Cockroach Control Chamber lures cockroaches into a baited black plastic station, where they are forced to rub shoulders with a fungus, *Metarhizium anisopliae*. This direct contact causes the fungus to release enzymes that weaken the cockroach's cuticle, allowing root-like hyphae to penetrate. The cockroach now becomes food for the fungus, which polishes off the innards after about two weeks. All the while, the cockroach spreads fungal spores among its harborage mates, who are then similarly devoured.

DEADLY WORMS

Just as insidious are the various nematode-packed chambers with futuristic trade names—Bio-Safe-N®, BioVector®, Helix®, and Magnet® to name a few. Cockroaches entering any of these products become infested with tiny roundworms. These internal parasites insinuate themselves through the cockroach's anus or spiracles. They deposit a bacterium that turns the cockroach's insides into fodder for the worm. Preliminary field tests have established that nematode chambers can be as effective as hydramethylnon bait stations. However, mass marketing of this promising technology has been hampered by the limited (one-year) shelf life of packaged stocks.

HI-TECH DEVICES

There's no evidence to suggest that electronic devices have any negative effects on cockroach populations. An evaluation of four ultrasonic pest-control

> To rid his apartment of roaches, Floridian Richard Dickerson set off nine insecticide foggers but forgot to douse the pilot light on his stove before leaving home. The resultant explosion brought down the ceiling, caved in a wall and shattered windows. "It worked much better then he expected," observed Tom Hurst of the Miami Beach Fire Department.

devices, conducted in 1983 by a team of University of Nebraska entomologists, concluded that "advertising claims that the sound would penetrate wooden doors, cabinets, and plaster walls (and control pests) should be considered as a gross exaggeration." During another set of laboratory tests, conducted on behalf of the United States Environmental Protection Agency (EPA), roaches actually became lodgers inside one of these insidious-looking but otherwise worthless black boxes.

From January 1979 to October 1980, a total of thirty-six enforcement actions were initiated by the EPA against the producers and distributors of electromagnetic pest control devices. These actions resulted in six civil complaints, twenty-one "stop sale, use or removal" orders, eight recalls, and one civil penalty—a fine of $1,250 assessed against Monty's Environmental Services, Inc., producers of a device called "The Eliminator." Acting under the assumption that "there's one born every minute" (suckers, that is—not roaches), bold entrepreneurs continue to manufacture and market equally useless electronic junk.

IGRS

Insect growth regulators (IGRs) are lab-created compounds that, when eaten, interfere with the cockroaches' manufacture of hormones for growth and reproduction. Two kinds are currently in use, but only one, juvenile hormone analogs (or juvenoids), is sold in the U.S. The other kind, called chitin synthesis inhibitors, have yet to be perfected for sale. Cockroach nymphs exposed to juvenoids develop into "adultoids"—intermediate forms with noticeable deformities, including curled or crinkled wings and unusually dark colored cuticles. Adultoids can go through all the motions of courtship, but are tragically unable to fully consummate their affairs.

A forgotten breakthrough

In 1980, Drs. Donald Cochran and Mary Ross uncovered a promising technique for controlling cockroach infestations. First, they subjected a laboratory culture of German cockroaches to high levels of gamma radiation, damaging the genetic material in the eggs and sperm of these animals. They then cross-bred these cockroaches and their mutant offspring. From these pairings, they were able to sire batches of sterile males, which could be released into healthy roach populations. Females that mated with these sterile males would produce egg capsules with defective embryos. The few healthy embryos in each batch were too weak to rupture their capsule's seam, and so they perished before they could hatch.

To test their newly developed technology, the two scientists released three batches of sterile males (a total of 424 insects) onto a cockroach-filled United States Navy vessel, conveniently drydocked at Norfolk, Virginia. These bred with the vessel's wild females, contaminating the population with their mutant genes for a period of four and a half months. Throughout this time, German cockroach populations declined in most parts of the ship.

Cochran and Ross were both guardedly optimistic about the results. They knew that the federal Department of Agriculture had already adopted a similar approach for eliminating the screwworm fly. By disseminating millions of sterile male screwworm flies over vast geographical areas, field technicians from this department were successful in eliminating this serious livestock parasite from nearly all of North America. Could an equally ambitious program lead to the decimation of German cockroach populations worldwide?

We may never know. According to Cochran, few homeowners or restaurateurs would ever agree to the release of additional cockroaches in their premises—in this particular instance, as many as nine times as many males from the special strain. As a result, the successful technique first tested on a ship docked in Norfolk, Virginia, may never be given a chance to prove itself on dry land.

The nuclear alternative

Why bother lobbing small, often ineffectual bombs at creatures with the strength of the Terminator and the persistence of the IRS? Wouldn't it be better to get it over with, once and for all? Put everything we've got into one, well-orchestrated nuclear attack?

Conventional wisdom suggests that even this desperate effort might fall short of snuffing out all the Blattaria on our planet. Data from laboratory experiments seem to support the notion that cockroaches would be capable of withstanding a good-sized thermonuclear blast.

The basic measure of an organism's ability to withstand radiation is the radiation absorbed dose, or rad—roughly equivalent to another measure of radioactivity, the roentgen. Humans who have absorbed radiation doses of around 300 rads will suffer some damage on a cellular level. A higher dosage (from 400 to 1,000 rads) over a period of two to three weeks will prove lethal.

Experiments conducted in the early 1960s by Drs. Cochran and Ross have shown that German cockroach adults and nymphs can survive radiation doses of 6,400 rads. Some of these specimens can tolerate even higher doses, succumbing only after absorbing 9,600 rads over thirty-five days. Such high tolerances to radiation would enable cockroaches to survive the atomic bombing of Hiroshima—a blast that bathed people as far as thirteen miles from Ground Zero with cumulative doses of around 1,200 rads.

That's the good news for cockroaches, and the few fatalistic souls who find cheer in the thought that insects will someday inherit the earth. The bad news, though, is that the devastation in Hiroshima was caused by a fifteen kiloton bomb—an explosive device seventy-six times less destructive than the one-megaton devices of today. Even these newer devices are considered small potatoes on a relative scale: our country's arsenal currently includes many nine-megaton bombs; the largest nuclear test, carried out by the former Soviet Union in 1961, had an explosive force of fifty-eight megatons. It is extremely unlikely that even the most firmly entrenched cockroach could withstand the radiation from such powerful blows.

Recently compiled projections indicate that a large-scale nuclear war would seal the fates of all but the most resilient life forms. In the first stages of such a conflagration, hundreds of large nuclear devices might be detonated within a matter of minutes. A ten thousand-megaton attack would create levels of radiation throughout North America averaging more than ten thousand rads. All mammals would be killed off, as would the majority of reptiles, amphibians, birds, and insects.

> "One of my native patients in S. Rhodesia always slept in the open, as he preferred to give up his hut to the cockroaches! He placed food there for them and slept outside undisturbed."
>
> —from "Hansen's Disease (Leprosy) and Cockroaches" by D.V. Moiser in the *East African Medical Journal* (1947)

Because cockroaches tend to hide during the day, some of these animals might be shielded from initial doses of radiation. But having escaped one lethal condition, they might soon find themselves faced with another: the thick clouds of soot and smoke from massive forest and brush fires. These clouds would block out all sunlight, causing a precipitous drop in temperatures worldwide. By some estimates, temperatures would average well below freezing, ushering in a life-terminating decade of nuclear winter. After a few decades, any cockroaches that could withstand such conditions would be confronted by a final, insurmountable obstacle, the depletion of the world's food supplies.

To make a long story short, humans *do* have the capacity to wipe out cockroaches, along with every other animal and plant. If any organisms *can* face up to our sinister instruments of destruction, it will be the bacteria, viruses, and a few primitive species of algae—all of which can withstand radiation doses in excess of ten thousand rads.

But who can say if our planet's evolutionary history won't repeat itself? Or if these meager remnants of the preapocalyptic past will evolve over billions of years, becoming progressively more complex, and eventually setting the scene from which higher forms of life will emerge? It's a good bet that the first land-dwelling animal to appear in this freshly reconstructed world will resemble a cockroach—one of the oldest and most successful beings on earth.

Resources

For More Information

ABOUT INSECTS

Borror, Donald J., Triplehorn, Charles A., and Johnson, Norman F., An Introduction to the Study of Insects. (New York: Harcourt Brace & Company, 1992)

Evans, Howard E., Life on a Little-Known (Planet. New York: Dutton, 1968)

Imes, Rick, The Practical Entomologist. (New York: Fireside Books/Simon and Schuster Inc., 1994)

ABOUT COCKROACHES (NONFICTION)

P. B. Cornwell, The American Cockroach: A Laboratory Insect and an Industrial Pest. (London: Hutchinson of London, 1976)

Guthrie, D. M., and Tindall, A.R., Biology of the Cockroach. (New York: St. Martin's Press, 1968)

Roth, Louis M., and Willis, Edwin R., The Biotic Associations of Cockroaches. (Washington: Smithsonian Miscellaneous Collections, volume 141, 1960)

Bell, W. J., and Adiyodi, K. G. (editors), The American Cockroach. (London, New York: Chapman and Hall, 1981)

Rust, Michael K., Owens, John M., and Reierson, Donald A. (editors) Understanding and Controlling the German Cockroach. (New York, Oxford: Oxford University Press, 1995)

ABOUT COCKROACHES (FICTION)

Harington, Donald, *The Cockroaches of Stay More.* (New York: Harcourt Brace Jovanovich, 1989)

Marquis, Don, *the lives and times of archy and mehitabel.* (New York: Doubleday & Company, Inc., 1940)

Weiss, Daniel Evan, *The Roaches Have No King.* (London, New York: High Risk Books, 1994)

ON THE INTERNET

Blattodea Culture Group home page http://www.ex.ac.uk/~gjlramel/bcg.html

Liberty (New Jersey) Science Center's "Yuckiest Site on the Internet" http://www.nj.com/yucky/roaches/index.html

To See Cockroaches...

AT MUSEUMS

Cockroach Hall of Fame
The Pest Shop, Inc.
2231-B West 15th Street
Plano, TX 75075
(214-519-0355)

The Insectarium
8046 Franklin Avenue
Philadelphia, PA 19136
(215-338-3000)

May Natural History Museum
of the Tropics
710 Rock Creek Canyon
Colorado Springs, CO 80926
(719-576-0450)

Dallas Museum of Natural History
P.O. Box 150349
Dallas, TX 75315-0349
(214-421-3466)

Ralph Parsons Insect Zoo
Natural History Museum
of Los Angeles County
900 Exposition Blvd
Los Angeles, CA 90007
(213-744-3466)

Otto Orkin Insect Zoo
MRC 158, Smithsonian Institution
Washington DC 20560
(202-357-1386)

AT ZOOS

Cincinnati Zoo Insectarium
3400 Vine Street
Cincinnati, OH 45220
(513-281-4700)

Fort Worth Zoological Park
   Insectarium
2727 Zoological Park Drive
Fort Worth, TX 76110-1787
(817-870-7057)

L'Insectarium de Montreal
4581 Rue Sherbrooke Est
Montreal, Quebec, Canada
(514-872-8753)

London Zoological Gardens Insect
   House
London Zoo, Regents Park
London, England
(071-722-3333)

Woodland Park Zoological Gardens
5500 Phinney Avenue North
Seattle, WA 98103
(206-684-4800)

To Catch, Buy, or Swap Your Own...

LIVE SPECIMENS

Blattodea Culture Group
c/o A. C. Barlow
71 Lower Ford Street,
Coventry CV1 59S, United Kingdom

Carolina Biological Supply Company
2700 York Road
Burlington, NC 27215
(1-800-537-7979)

Connecticut Valley Biological
P.O. Box 326
Southampton, MA 01073
(1-800-628-7748)

COLLECTING AND HUSBANDRY SUPPLIES

BioQuip Products, Inc.
17803 LaSalle Avenue
Gardena, CA 90248
(213-324-0620)

Wards Natural Science Establishment
P.O. Box 92912
Rochester, NY 14623
(1-800-962-2660)

173

RUBBER ROACHES & ASSORTED COCKROACH-ABILIA

Archie McPhee
P.O. Box 30852
Seattle, WA 98103
(206-783-2344)

Young Entomologist's Society, Inc.
1915 Peggy Place
Lansing, MI 48910-2553
(517-887-0499)

Handpuppet available from:
Puppets on the Pier
(800) 443-4463
http:/www. americandreams.com/puppets

Roach costume patterns:
Foam Domes
5335 Starview Lane
Prior Lake, MN 55372-9600
(612) 447-0906

To Control Cockroaches

WITH LEAST-TOXIC ALTERNATIVES

Bio-Integral Resources Center
P.O. Box 7414
Berkeley, CA 94707
(510-524-2567)

North American Hedgehog Association
 P.O. Box 122, Nogal
New Mexico 88341

Sure-Fire Products
213 SW Columbia Street
Bend, OR 97702
(541-388-3688)

USING STRONGER STUFF

National Pest Control Association
8100 Oak Street
Dunn Loring, VA 22027
(703-573-8330)

Pest Control Magazine
Edgell Communications, Inc.
7500 Old Oak Boulevard
Cleveland, OH 44130

Index

A

Activity patterns, 83, 90–91
Aggression, 17–18
Airplanes, 41
Albinos, 77, 79
All-American Trot, 85–86
Allergies, 53–55
American cockroach. *See Periplaneta americana*
Antennae fencing, 65
Ants, 26–27
Aptera cingulata (Cape Mountain roach), 78
Aquatic species, 89–90
Arboreal species, 27–28
Archy the cockroach, x, 84, 122, 125–27, 130
Arenivaga sp. (desert cockroach), 7, 46
Arenivaga investigata (sand cockroach), 107
Armstrong, Louis, 50, 60, 61
Art
  cockroaches eating, 100
  cockroaches in, 117–21
Asian cockroach. *See Blattella asahinai*
Aspiduchus cavernicola (Tuna Cave cockroach), 42–43
Asthma, 53, 54
Attaphila fungicola, 26–27
Australian cockroach. *See Periplaneta australasiae*

B

Bacillus thuringensis, 166
Banded wood cockroach.
  See Parcoblatta zebra

Bicolored cockroach. *See Pycnoscelus surinamensis*
Biosphere 2, 46
Birth, 75, 76
Blaberidae, 6–7
Blaberus craniifer, 9
Blaberus discoidalis, 137
Blaberus giganteus (giant cockroach), 6–7, 77
Blaberus sp., 68, 71
Blatta orientalis (Oriental cockroach), 23, 38, 49, 50
Blattaria, 5
Blattella asahinai (Asian cockroach), 88–89
Blattella germanica (German cockroach), 9, 23, 38, 75, 77, 79–81, 83, 84, 88–89, 102–3
Blattella vaga (field cockroach), 7
Blattellidae, 7, 75
Blattidae, 6, 74
Blattodea Culture Group, 137, 173
Body structure, 4, 11–15
Boll's wood cockroach. *See Parcoblatta bolliana*
Book of the Dead, 149
Books, eating, 99–100
Boric acid, 163
Boscarino, Richard, 119
Breathed, Berke, 122
Broad Keys cockroach. *See Hemiblabera tenebricosa*
Broad wood cockroach.
  See Parcoblatta lata
Brownbanded cockroach.
  See Suppela longipalpa

Brown cockroach. *See Periplaneta brunnea*
Brown hooded cockroach. *See Cryptocercus punctulatus*
Buhach Company, 160–61
Byrsotria fumigata (Cuban burrowing cockroach), 78

C

Camouflage, 112
Cannibalism, 102–3, 144
Canopy cockroach. *See Paratropes bilunata*
Cape Mountain roach.
  See Aptera cingulata
Captain Cockroach, 122
Carolina Biological Supply Company, 47, 137, 173
Cartoons, 123–25
Cats, 111–12
Caudell's wood cockroach.
  See Parcoblatta caudelli
Caves, 28
Centipedes, 110–11
Cinereous cockroach.
  See Nauphoeta cinerea
Cities, cockroaches in, 29
Cleaning habits, 91–92
Cockroaches of Stay More, The, 128
Cockroach Hall of Fame, 26
"Cockroach," origin of, 10
Cockroach That Ate Cincinnati, The, 131–32
Cocky Cockroach, 123
Collecting cockroaches, 137–42, 146–47, 173
Combat bait stations, 162

Comics, 122–23
Congress, cockroaches in, 31
Connecticut Valley Biological, 137, 173
Control methods
  environment alteration, 154–56
  geckos, 158–59
  hedgehogs, 153–54
  herbal, 159–61
  historical, 149–51
  information sources, 174
  insecticides, 164–65
  new, 94–95, 166–67
  nuclear radiation, 168–70
  poisoned baits, 161–63
  powders, 163
  ridiculous, 152
  on ships, 153
  sterilization, 167–68
  traps, 156–58
Costumes, 59, 174
Courtship rituals, 65–66, 69, 70, 71
Creepshow, 28, 58
Cryptocercidae, 7
Cryptocercus clevelandi, 7, 21
Cryptocercus primarius, 7
Cryptocercus punctulatus (brown hooded cockroach), 7, 35–36
Cryptocercus relictus, 7
Cuban burrowing cockroach. *See Byrsotria fumigata*
Cypress cockroach. *See Diploptera punctata*

D

Dances, 132, 134
Death pose, 165
Defensive mechanisms, 113
Desert cockroach. *See Arenivaga sp.*
Desert wood cockroach. *See Parcoblatta desertae*
Diet. *See* Food
Diploptera punctata (Pacific beetle cockroach), 9, 75, 156

Disease, 51–52
Distribution
  in the United States, 20–21
  in the world, 19–20
Domino Chance, 122–23
Drake, Sir Francis, 39
Dusky cockroach. *See Ecotobius lapponicus*

E

Ecotobius lapponicus (dusky cockroach), 77
Egg capsules. See Ootheca
Electromagnetic pest control devices, 167
El Malefico de la Mariposa, 130–31
Embryonic molt, 76
Endangered species, 42–43
Entomophobia, 56–57
Epilampra abdomennigrum, 90
Epilampra sp., 45
Epilampra wheeleri, 107
Epilamprinidae, 89
Eurycotis decipiens (Trinidad zebra cockroach), 145
Eurycotis floridana (stinking cockroach), 113
Eurycotis improcera, 107
Evolution, 33–36

F

Families, taxonomic, 6–7
Field cockroach. *See Blattella vaga*
Florida beetle cockroach. *See Plectoptera poeyi*
Flying species, 4, 87–89
Food
  for cockroaches, 97–104, 141–42, 154
  cockroaches as, 47–48
Freleng, Friz, 123
Frischman, Austin, 30
Fulvous wood cockroach. *See Parcoblatta fulvescens*

G

Garcia-Lorca, Federico, 130
Geckos, 158–59
German cockroach. *See Blattella germanica*
Giant burrowing cockroach. *See Macropanesthia rhinoceros*
Giant cockroach. *See Blaberus giganteus*
Glucose aversion, 162
Gromphadorhina portentosa (Madagascan hissing cockroach), 9, 46, 70, 137, 140

H

Harborages, 92–94, 96, 155
Harington, Donald, 128
Harlequin cockroach. *See Neostylopyga rhombifolia*
Hedgehogs, 153–54
Hemiblabera tenebricosa (broad Keys cockroach), 21
Herriman, George, 122
Holocompsa nitidula (small hairy cockroach), 21

I

IGRs. *See* Insect growth regulators
Infestation levels, estimating, 30, 158
Insectarium, The, 145
Insect growth regulators, 167
Insecticides, 29, 164–65
Insects
  ancient, 33–36
  classification of, 5
  number of species, 3
Intelligence, 15–16
Internet, 172
Ischnoptera rufa, 107

J

Joe's Apartment, 58

K

Kafka, Franz, 129–30
King, Albert, 131

L

Laboratory experiments, 46–47, 143–44
La Cucaracha, 59–61
Lady in Red, The, 123
Lamb, Gina, 121
Larsen, Gary, 122
Latiblatta lucifrons (pale-headed cockroach), 46
Legal issues, 53
Lenagh, Kevin, 122–23
Life span, 16
Literature, cockroaches in, 125–31
Live-bearing species, 75
Locomotion, 84

M

Macropanesthia rhinoceros (giant burrowing cockroach), 25
Madagascan hissing cockroach. *See Gromphadorhina portentosa*
Madeira cockroach. *See Rhyparobia maderae*
Magnifiers, 142
Marquis, Don, 125–26
McPhee, Archie, 59, 174
Medicine, 49–50
Megaloblatta blaberoides, 24
Megaloblatta longipennis, 24
Merian, Maria Sibylla, 117, 119
Metamorphosis, The, 129–30
Metarhizium anisopliae, 166
Mice, 111
Migrations, 81–82
Molting, 77–78
Mouffet, Thomas, 37–38, 39, 150
Movies, 58
Museum exhibits, 145, 172

N

Names
changing, 9
historical, 9–10
scientific, 8–9
Nauphoeta cinerea (cinereous cockroach), 9, 69
Nematodes, 166
Neostylopyga rhombifolia (harlequin cockroach), 40
Nocticola caeca, 28
Nuclear radiation, 168–70
Nymphs, 76–79

O

Ocampo, Manuel, 119
Odors, 50
Ootheca, 6, 73–75
Oriental cockroach. *See Blatta orientalis*
Ovoviviparity, false, 75, 78

P

Pacific beetle cockroach. *See Diploptera punctata*
Pale-headed cockroach. *See Latiblatta lucifrons*
Palmetto bug. *See Periplaneta americana*
Panchlora nivea, 107
Paratropes bilunata (canopy cockroach), 46
Parcoblatta americana (western wood cockroach), 21
Parcoblatta bolliana (Boll's wood cockroach), 21
Parcoblatta caudelli (Caudell's wood cockroach), 21
Parcoblatta desertae (desert wood cockroach), 21
Parcoblatta divisa (southern wood cockroach), 21
Parcoblatta fulvescens (fulvous wood cockroach), 21
Parcoblatta lata (broad wood cockroach), 21

Parcoblatta pensylvanica (Pennsylvania wood cockroach), 21, 22
Parcoblatta uhleriana (Uhler's wood cockroach), 21
Parcoblatta virginica (Virginia wood cockroach), 21
Parcoblatta zebra (banded wood cockroach), 21
Parthenogenesis, 72
Pennsylvania wood cockroach. *See Parcoblatta pensylvanica*
Peridomestic cockroaches, 22
Periplaneta americana (American cockroach), 6, 22, 68, 74, 83, 84, 102–3, 107
Periplaneta australasiae (Australian cockroach), 6, 46, 107
Periplaneta brunnea (brown cockroach), 6, 9
Periplaneta fuliginosa (smokeybrown cockroach), 23
Perisphaerus semilunatus, 79
Pets, cockroaches as, 135–42
Pheromones, 35, 65–66, 79, 94
Plectoptera poeyi (Florida beetle cockroach), 21
Poisoned baits, 161–63
Pollination, 46
Polyphagidae, 7
Population
in cities, 29
estimating, 30, 158
growth, 79–81
Powders
boric acid, 163
diatomaceous earth, 163
pyrethrin, 160–61
silica gel, 163
Predators, 105–12, 153–54, 158–59
Preserved cockroaches, 146–47
Prosoplecta sp., 112
Protein content, 45–46
Prussian cockroach. *See Blattella germanica*

Pseudofemale behavior, 71
Psychological studies, 55–57
Pulvis tarakanae, 49
Puppets, 59, 174
Pycnoscelus surinamensis
    (Surinam cockroach), 46,
    71–72, 107
Pyrethrin, 160–61

R
Races, 85–86
Raid Research Institute, 144
Regeneration, 78
Reproduction, 14, 65–66, 68,
    71–75
Resistant roaches, 162
Rhyniella praecursor, 35
Rhyparobia maderae (Madeira
    cockroach), 46, 69, 75
"Roach," origin of, 10
Roacharium, 140–41
Roach Dance, 132, 134
Roaches Have No King, The,
    128
Robots, 86–87
Roches, the, 132
Rubber roaches, 59, 174

S
S. C. Johnson Corporation, 31,
    144
Sand cockroach. See Arenivega
    investigata
Schultesia lampyridiformis, 112
Ships, 38–41, 153
Sim, Dave, 122
Singing. See Stridulation
Size, largest, 24–26
Small hairy cockroach.
    See Holocompsa nitidula
Smokeybrown cockroach.
    See Periplaneta fuliginosa
Songs, 59–61, 131–32
Sounds
  hissing, 70
  stridulation, 68–69
  tapping, 70

Southern wood cockroach.
    See Parcoblatta divisa
Spaceships, 42
Species
  in the United States, 20–23
  in the world, 19
Speed, 84, 113
Spiders, 109–10
Sterilization, 167–68
Stinking cockroach. *See Euryco-
tis floridana*
Stridulation, 68–69
Suppela longipalpa (brown-
    banded cockroach), 23, 84
Suppliers, 137, 173
Surinam cockroach. *See Pyc-
noscelus surinamensis*

T
Taxonomy, 5–9
Terry, Paul, 123
Thigmotaxis, 92
Tractor pulls, 86
Traps, 140, 151, 156–58
Trinidad zebra cockroach.
    See Eurycotis decipiens
Tuna Cave cockroach.
    See Aspiduchus cavernicola
TV roaches. *See Suppela longi-
palpa*
Twilight of the Cockroaches,
    124–25

U
Uhler's wood cockroach.
    See Parcoblatta uhleriana
Ultrasonic pest control devices,
    166–67
Urates, 68, 72–73

V
Villa, Francisco "Pancho," 60
Virginia wood cockroach.
    See Parcoblatta virginica

W
Wasps, 108–9
Water requirements, 103–4, 155
Waters, John, 132
Webb, Dick, 119, 121
Weiss, Daniel Evan, 128
Western wood cockroach. *See
    Parcoblatta americana*
White, E. B., 126
White, Stanley, 123

X
Xestoblatta hamata, 72
Xestoblatta sp., 98

Y
Yoshida, Hiroaki, 124–25

Z
Zapper, 94–95
Zoo exhibits, 108, 144–45, 173